DIONEX

Handbook of Ion Chromatography

Handbook of Ion Chromatography

Authored by: Joachim Weiss, Germany

Edited by: Edward L. Johnson
Dionex Corporation
Sunnyvale, California

Aminex is a registered trademark of Bio-Rad Labs, Richmond, California. **AutoIon** and **HiPhi** are registered trademarks of Dionex Corporation, Sunnyvale, California. **Dowex** is a registered trademark of Dow Chemical Company, Midland, Michigan. **Teflon, Tefzel,** and **Zorbax** are registered trademarks of E.I. DuPont de Nemours and Company, Wilmington, Delaware. **LiChrosorb** is a registered trademark of E. Merck Darmstadt, Germany. **Kel-F** is a registered trademark of Minnesota Mining and Manufacturing Company, St. Paul, Minnesota. **Dequest** is a registered trademark of Monsanto Company, St. Louis, Missouri. **Hyamine** is a registered trademark of Rohm and Haas, Philadelphia, Pennsylvania.

Vydac is a trademark of Separations Groups, Hesperia, California.

This book is published by **Dionex Corporation, Sunnyvale, California.**

Design and composition by The Bookmaker, San Jose, California. Phototypesetting by ReederType, Fremont, California. Printing by Bofors, Inc., San Mateo, California. Cover art by Deborah Knodel, Sunnyvale, California.

ISBN 0-961-7173-0-0
ABCDEFGHIJ-BL-89876
LPN 32769 7/86

Foreword

Without a doubt, this book is the best book to date dealing with Ion Chromatography. The author has complemented much of the available literature with additional data derived from his own work. This new data is used to enhance and clarify a variety of points throughout the book.

As much as possible, the English version remains faithful to the original German monograph. The forewords remain in German since a translation would detract from their purpose. It has been my pleasure to play a small part in bringing this important work to the English reading market.

Edward L. Johnson
June 9, 1986

Geleitwort

Die Ionen-Chromatographie in ihrer jetzigen technisch weit entwickelten Form ist mit Sicherheit einer der bedeutendsten Beiträge zur instrumentellen Analytik von Ionen innerhalb des letzten Jahrzehnts.

Man denke etwa zehn Jahre zurück und an die Aufgabe, aus einer Laboratoriumsprobe ein gutes Dutzend Anionen zu bestimmen, und das womöglich im ppb-Bereich. Heute ist ein solches Problem rasch gelöst, wenn man einen ordentlichen Ionenchromatographen und die richtigen Trennsäulen besitzt.

Der Ursachen für diese auf den ersten Blick verblüffende analytische Leistungsfähigkeit der Ionen-Chromatographie in einer schier unglaublichen Vielfalt von Anwendungsbereichen sind zwei. Zum einen handelt es sich um eine HPLC-Methode im besten Sinne des Wortes, der hohe chromatographische Effizienz und Selektivität zu eigen sind, zum anderen hat man Detektionssysteme entwickelt, die Analysen im unteren ppb-Bereich erlauben. Die Feststellung, daß die moderne Ionen-Chromatographie eine HPLC-Methode ist, stößt bisweilen auf Widerspruch. Zugegebenermaßen sind die Trennsäulen der Ionen-Chromatographie meist etwas anders aufgebaut, doch ist die Vielfalt der HPLC-Methoden (ohne Präfix Ion) schon so weit, daß auch die HPIC darin keinen Fremdkörper darstellt.

Der Verfasser dieses Buches kommt aus einem weitestgehend chromatographisch arbeitenden Labor und hat - wie hier glänzend demonstriert - keine Schwierigkeiten, sich auch auf dem Feld ionischer Soluten umzutun. Daß für manche Ionentrennung noch keine perfekte theoretische Erklärung vorliegt, hat den Verfasser nicht abgeschreckt, kannte er diesen Zustand doch schon aus dem nichtionischen Bereich. Die praktische Anwendbarkeit einer Methode ist in der Analytik nach wie vor wichtiger als ein vollständiger theoretischer Unterbau, wobei letzterer je nach Entwicklungsstand der Methode meist nachwächst.

Zur Zeit befindet sich die Chromatographie ionischer Soluten in einer so lebhaften Entwicklung, daß ein lückenloses Verfolgen aller einschlägigen Publikationen schon mühsam wird. Es war daher im höchsten Maße zu begrüßen, daß ein Buch greifbar wurde, das den aktuellen Stand der Ionen-Chromatographie zuverlässig und kritisch darstellt.

Das Buch von Joachim Weiß tut dies in umfassender und solider Form. Die praktischen Aspekte sind so detailliert beschrieben, wie man es sich nur wünschen kann. Die Darstellung theoretischer Zusammenhänge ist hervorragend gelungen und beschränkt sich auf Themen, die nach dem heutigen Kenntnisstand guten Gewissens vertretbar sind.

Dabei ist ein echtes Handbuch entstanden, dem meine besten Wünsche mit auf den Weg zu geben, mir eine große Freude ist.

Hermann J. Möckel

Januar 1985

Hahn-Meitner-Institut für Kernforschung

Berlin-Wannsee

Vorwort des Verfassers

Das Ziel des vorliegenden „Handbuches" ist es, grundsätzliche Aspekte der Ionen-Chromatographie aufzuzeigen und dem Leser diese in leicht verständlicher Form zugänglich zu machen. Daher sollen neben den theoretischen Grundlagen der Ionen-Chromatographie vor allem die analytischen Anwendungsmöglichkeiten dieser Methode in ihrer gesamten Breite diskutiert werden.

Die rasche Entwicklung der Ionen-Chromatographie seit ihrer Einführung im Jahre 1975 und der Mangel an deutschsprachiger Literatur zu diesem Thema gaben den Anlaß zur Entstehung dieses Handbuches. Es soll all denen als Leitfaden dienen, die bereits mit dieser Methode arbeiten oder sich für ihre Anwendungen interessieren. Darüber hinaus ist es als Grundlage für die Ionen-Chromatographie-Kurse gedacht, die Dionex regelmäßig in Weiterstadt durchführt.

Um das Studium der weiterführenden angelsächsischen Literatur zu erleichtern, wurde bei Fachausdrücken der englisch/amerikanische Begriff beigefügt. Dimensionen und Maße sind im metrischen System angegeben. Angelsächsische Fachausdrücke, für die es kein direktes deutsches Äquivalent gibt, wurden apostrophiert übernommen.

All denen, die am Zustandekommen dieses Handbuches beteiligt waren, sei an dieser Stelle herzlichst gedankt. Mein besonderer Dank gilt meiner Frau Ursula für ihre stete Unterstützung während der Erstellung des Manuskriptes.

Für Hinweise auf Fehler, die trotz aller Sorgfalt unentdeckt geblieben sein könnten, wäre der Verfasser ebenso dankbar wie für Verbesserungsvorschläge, die einer späteren Auflage zugute kommen würden.

Joachim Weiß — Weiterstadt, im März 1985

Contents

Chapter 3 — Ion Exchange Chromatography (HPIC) 19

Chapter 4 — Ion Exclusion Chromatography (HPICE) 92

Chapter 5 — Ion Pair Chromatography (MPIC) 114

Chapter 6 — Modes of Detection 141

Chapter 7 — Quantitative Analysis 160

Chapter 8 — Instrumentation 177

Chapter 9 — Applications 185

1 Introduction

1.1 HISTORY

The general term "chromatography" is used to denote a number of physico-chemical separation techniques based on the distribution of solutes between a mobile and a stationary phase. The classification of the various types of chromatography is based on the nature of the two phases involved.

The discovery of chromatography has been credited to Twsett [1,2], a Russian who, in 1903, succeeded not only in separating leaf pigments using a solid polar stationary phase, but also in interpreting the chromatographic process. In the following years, chromatography was restricted to the use of a solid stationary phase and a liquid mobile phase (LSC = Liquid Solid Chromatography). In 1938, Izmailov and Schraiber [3] laid the foundation for thin layer chromatography (TLC) which was later refined by Stahl [4,5] in 1958. In their remarkable work of 1941, Martin and Synge [6] applied the concept of theoretical plates, from distillation technology, as a formal measurement of the efficiency of the chromatographic process. This concept not only revolutionized liquid chromatography (LC) techniques, but set the stage for the development of gas and paper chromatography.

1952 saw the publication of the first gas chromatographic work by Martin and James [7] initiating an extremely rapid development of this technique.

High Performance Liquid Chromatography (HPLC), derived from classical column chromatography, has today become the most important of modern instrumental analytical techniques. The breakthrough in HPLC was achieved by the development of very efficient separators. These new separators utilized small particles ($\approx$10 μm) which required the use of high pumping pressures. The development of HPLC and the theoretical understanding of the separation processes involved have been, in particular, dependent on the fundamental studies by Horvath [8], Knox [9], Scott [10], Snyder [11], Guiochon [12], Möckel [13], et al.

Ion Chromatography (IC), as a novel analytical method, was introduced in 1975 by Small, Stevens and Baumann [14]. The technique has developed within a very short time, from a new way of detecting a few inorganic anions to a versatile analytical technique for ionic species of all kinds. Modern IC now involves unique combinations of a number of separation systems with appropriate detectors. This rapid development has been due to a significant improvement in the understanding about the ion exchange materials employed and of the separation processes occurring.

1.2 TYPES OF ION CHROMATOGRAPHY

This book deals exclusively with the various separation techniques grouped under the headling "Ion Chromatography." Modern IC, as a special type of liquid chromatography, is based on three separation techniques which provide the basis for the nomenclature.

Ion Exchange Chromatography (HPIC)
(High Performance Ion Chromatography)

The essential principle of this technique is an ion exchange process between the mobile phase and the exchange groups covalently bound to the stationary phase. For hydrophobic ions, in particular, adsorption processes are also involved. The stationary phase is typically polystyrene based resin which has been crosslinked with divinyl benzene. Ion exchange chromatography can be used for separating both inorganic and organic anions and cations. For the analysis of anions, the exchange function usually is a quarternary ammonium group; whereas for the analysis of cations, a sulphonate group is most often employed. A more detailed discussion of this technique is given in Chapter 3.

Ion Exclusion Chromatography (HPICE)
(High Performance Ion Chromatography Exclusion)

Ion exclusion chromatography involves Donnan exclusion, steric exclusion, and adsorption processes. The stationary phase is a totally sulphonated, high capacity cation exchange resin derived from a polystyrene/divinyl benzene copolymer. Ion exclusion chromatography is especially useful for separating weak organic acids from totally dissociated acids; the latter being eluted as a single peak at a position corresponding to the void volume of the separator. Additionally, this separation technique can be used for the determination of carbonate and borate. A detailed discussion is given in Chapter 4.

Ion Pair Chromatography (MPIC)
(Mobile Phase Ion Chromatography)

The dominant process involved in ion pair chromatography is adsorption. Using MPIC, it is possible to separate metal complexes and surfactant anions and cations. Thus, the selectivity of the column is determined by the mobile phase. Besides organic and inorganic modifiers, an ion pair reagent is added to the eluant. The stationary phase is composed of a neutral, non-polar, macroporous polystyrene/divinyl benzene based resin. A more detailed description of this technique is given in Chapter 5.

1.3 THE ION CHROMATOGRAPHIC SYSTEM

The basic components of a modern ion chromatograph as developed by Dionex are shown as a block diagram in Figure 1-1.

The mobile phase is pumped through the chromatographic system by a double reciprocating pump. A pulse-free flow, necessary for both the sensitive UV/Vis and amperometric detectors, is ensured by a complex electronic control of the pump.

The samples are introduced into the system via a loop injector which is shown schematically in Figure 1-2.

The two outlets of the sample loop, which is loaded at atmospheric pressure, are joined by a three-way valve. After the valve is switched, the sample is transported onto the column by the mobile phase.

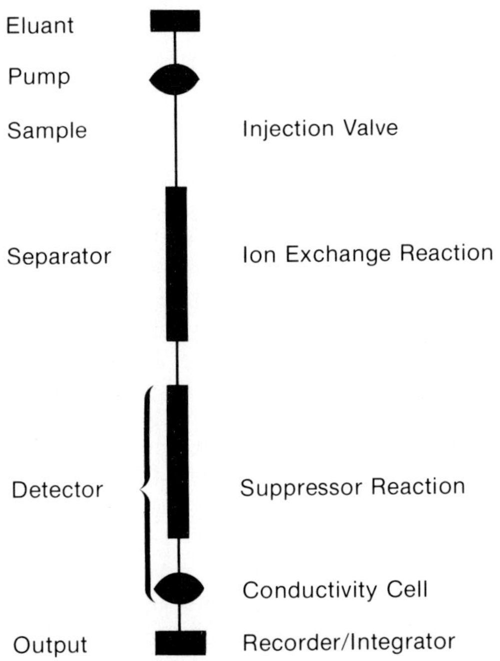

FIGURE 1-1. Basic components of an ion chromatograph.

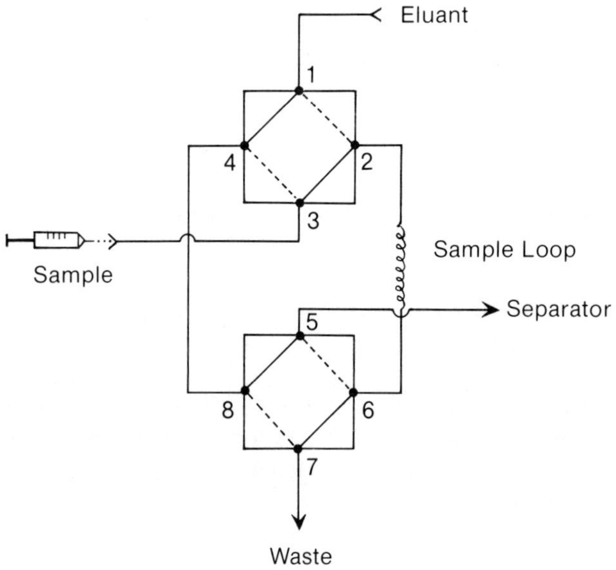

FIGURE 1-2.
The loop injector.

The most important part of the chromatograph is the analytical column. The selection of a suitable stationary phase (see Chapter 1, Section 5), as well as of the appropriate chromatographic parameters, determines the quality of the analysis. The column bodies are made of inert materials and are generally operated at room temperature. Only in a few cases, such as the analysis of carbohydrates and amino acids, is it necessary to thermostat the columns at elevated temperatures.

In order to assess the performance of the detector (which serves to identify and quantify the species being analyzed) the following criteria are examined:

- Linearity
- Resolution
- Noise

The most widely used detector in IC is the conductivity detector. This detection system usually includes a suppressor device that serves to chemically reduce the background conductivity of the eluant which, at the same time, converts the species of interest into a more conductive form. In addition to the conductivity detector, UV/Vis, amperometric, and fluorescence detectors are sometimes also employed. These are described in more detail in Chapter 8.

The chromatogram is obtained via a recorder or computer. Quantitative results are derived from a calculation of the peak areas, or peak heights, that are proportionate to the concentration of the species to be determined. Simplification of the data reduction task is frequently accomplished via digital integrators or personal computers.

1.4 ADVANTAGES OF ION CHROMATOGRAPHY

The determination of ionic compounds in solution is a classical analytical problem and a wide variety of techniques have been used. Although a number of techniques for the analysis of cations has been available for some time (e.g., AAS, AES, ICP), the analysis of anions has suffered from a deficiency of highly sensitive techniques. In comparison with conventional wet chemical techniques (such as photometry, gravimetry, turbidimetry, and colorimetry which tend to be labor intensive, time consuming, and, to some extent, error prone), ion chromatography offers the following advantages:

- Speed
- Sensitivity
- Selectivity
- Simultaneous Analysis

1. Speed

Due to increasing production costs for high quality goods and increasing interest in environmental protection, the speed of analytical procedures is becoming more and more important as the number of samples for analysis rapidly increases.

The introduction of a new generation of high performance separators for ion exchange and ion exclusion chromatography has reduced the analysis time to less than 15 minutes for most common anions. For example, the baseline resolved separation of seven inorganic anions can now be achieved in ca. 6 minutes. Quantitative results can thus be obtained in a fraction of the time needed for the traditional wet chemical techniques with a consequent increase in the rate of sample throughput.

2. Sensitivity

Since the introduction of microprocessor technology in conductivity detection, it has become possible to identify ions in the middle and lower ppb concentration ranges without preconcentrating the samples. Hence, direct injection techniques have a detection limit for simple anions of ca. 10 ppb. The absolute amount of sample injected is, thus, in the order of nanograms. Sample enrichment on a concentrator column permits the analysis of even carefully deionized water for the presence of chloride, sulphate, and/or sodium ions. Using this technique, the detection limit can be reduced to around 0.010 ppb. For amino acid analysis (using a fluorescence detector after postcolumn reaction with o-phthalaldehyde), detection limits in the picomole range can be achieved.

3. Selectivity

The selectivity of ion chromatographic procedures for analyzing inorganic and organic anions and cations is ensured by the choice of suitable separation and detection systems. In this respect, the suppressor technique, by which the counterions are exchanged for H^+ (or OH^-), is of paramount importance. With the aid of derivatization of the chromatographically separated species, more recent developments in ion chromatography demonstrate that heavy and transition metals, polyvalent anions, and silicate can be determined with improved selectivity. Examples of this technique explain why the sample preparation in ion chromatography involves, in many cases, only a simple dilution and/or filtration. Furthermore, the high degree of selectivity simplifies the identification of unknown sample components.

4. Simultaneous Analysis

A major advantage of ion chromatography—also in comparison with other instrumental techniques, such as potentiometry and AAS—is the possibility of simultaneously determining a number of individual components in a sample. Thus, in a very short time, one obtains an anion or cation profile that provides information about the composition of the sample and avoids the necessity of time consuming tests.

A unique advantage of IC, compared to HPLC, is the stability of the stationary phases employed. This is due to the kind of materials from which they are prepared. In contrast to silica-based separators used in HPLC, polystyrene/divinyl benzene copolymers are used as support material for IC separators. The improved pH stability of these materials allows strong acids and bases to be used as eluants—a prerequisite for the versatility of this technique. Furthermore, the stability of the separators implies that they are less affected by complex matrices. Thus, an analysis of the iodide content of milk can be carried out simply by dilution and direct injection.

1.5 CHOICE OF COLUMNS AND DETECTORS

As already discussed, three separation techniques are included under the general heading "Ion Chromatography." Therefore, in this section, the criteria involved in selecting the appropriate stationary phases and detectors is discussed.

In general, the analytical chemist has a knowledge of the nature (inorganic or organic), the hydrophobicity, and the pK_a values of the ions to be determined. With this information and the flow diagram in Table 1-1, the choice of a suitable stationary phase and detection system for a given analytical problem is a straightforward procedure. However, it should be pointed out that, not uncommonly, several techniques can be suitable for a single analytical problem. In order to emphasize this point, three examples are described below.

TABLE 1-1: Flow diagram for selecting the appropriate separator and detector for a particular analytical problem.

Anions	Hydrophobic Ions, I^-, ClO_4^-, Phenolate, Carboxylic Acids and Sulfonates $>C_4$, etc.	$pK_a>7$	Organic Ions, Phenols, etc.	MPIC	Amperometric or UV/Vis Detection
			Carbohydrates	HPIC	Pulsed Amperometry
			Inorganic Ions, Arsenite, etc.	HPIC	Amperometric or UV/Vis Detection
		$pK_a<7$	Organic and Inorganic Ions ClO_4^- Sulfonate$>C_4$	HPIC or MPIC	Conductivity Detection
	Hydrophilic Ions, F^-, Cl^-, SO_4^{2-}, Carboxylic Acids and Sulfonates $<C_4$, etc.	$pK_a>7$	Organic Ions, Mercaptans	HPIC or MPIC	Amperometric Detection
			Inorganic Ions, HS^-, CN^-, etc.	HPIC	Amperometric or UV/Vis Detection
			Alcohols	HPICE	Pulsed Amperometry
		$pK_a<7$	Organic Ion $<C_4$	HPICE or HPICE/HPIC or MPIC	Conductivity or UV/Vis Detection
			Chelants	HPIC	UV/Vis Detection with Postcolumn Reaction
			Inorganic Ions F^-, Cl^-, SO_4^{2-}, etc.	HPIC or HPICE/HPIC	Conductivity or Amperometric Detection
Cations	Hydrophobic Ions, Aryl, Alkylamines $>C_3$			MPIC	Conductivity Detection for $pK_a<7$ or UV/Vis Detection
	Hydrophilic Ions, Na^+, Mg^{2+}, NH_4^+, Heavy Metals, Amines $<C_3$, Amino Acids		Amines$<C_3$	MPIC or HPIC	Conductivity Detection for $pK_a<7$ or UV/Vis Detection
			Transition and Heavy Metals	HPIC	UV/Vis Detection with Postcolumn Reaction
			Alkali and Alkaline Earths	HPIC	Conductivity Detection
			Amino Acids	HPIC	Fluorescence Detection with Postcolumn Reaction

Problem 1: *The analysis of a variety of sulfur-containing species in the circulating water within a flue gas desulfurization plant.*

According to Table 1-1, hydrophilic ions (such as sulphite and sulphate with $pK_a < 7$) can be separated by HPIC and detected by suppressed conductivity detection. The more hydrophobic dithionate, also present in these solutions, can only be separated by ion pair chromatography (MPIC). The detection system of choice is, again, suppressed conductivity detection.

Problem 2: *The separation and quantification of anionic surfactants.*

Such surface-active compounds are usually strong organic acids with a hydrophobic character. They can be separated by MPIC and detected via their conductivity. If, however, the surfactant has an aromatic moiety, UV/Vis detection is often more sensitive than suppressed conductivity detection.

Problem 3: *The determination of heavy metals in baby food.*

Depending on the degree of selectivity desired, heavy metals (such as hydrophilic cations) can be separated by either HPIC or MPIC. In both cases, they can be detected utilizing UV/Vis absorption after postcolumn derivatization of the column effluent.

Additional details that will help to select the most suitable technique for the problem at hand are given in Chapters 3, 4, and 5.

LITERATURE

1. M. Tswett, Trav. Soc. Nat. Var. **14**, 1903 (1903).
2. M. Tswett, Ber. Deut. Botan. Ges. **24**, 385 (1906).
3. N.A. Izmailov and M.S. Schraiber, Farmatsiya **3**, (1938).
4. E. Stahl, Pharmazie **11**, 633 (1956).
5. E. Stahl, Chemiker Ztg. **82**, 323 (1958).
6. A.J.P. Martin and R.L. Synge, Biochem. J. **35**, 91, 1358 (1941).
7. A.T. James and A.J.P. Martin, Analyst **77**, 915 (1952).
8. C. Horvath, W. Melander and I. Molnar, J. of Chromatogr. **125**, 129 (1976).
9. J.H. Knox, "Theory of HPLC, Part II: Solute Interactions with the Mobile Phase and Stationary Phases in Liquid Chromatography," C.F. Simpson (editor), "Practical High Performance Liquid Chromatography," Heydon and Son Ltd. (1976).
10. R.P.W. Scott, "Theory of HPLC, Part II: Solute Interactions with the Mobile Phase and Stationary Phases in Liquid Chromatography," C.F. Simpson (editor), "Practical High Performance Liquid Chromatography," Heydon and Son Ltd. (1976).
11. L.R. Snyder, Chromatogr. Rev. **7**, 1 (1965).
12. G. Guiochon, "Optimazation in Liquid Chromatography," C. Horvath (editor), "High Performance Liquid Chromatography, Vol. 2," Academic Press Inc. New York (1980).
13. H.J. Mockel, Lecture: "Instrumentelle Analytik I," TU Berlin 1974-1984.
14. H. Small, T.S. Stevens and W.C. Baumann, Anal. Chem. **47**, 1801 (1975).

2 Formal Theory of the Chromatographic Process

2.1 CHROMATOGRAPHIC PARAMETERS

To fully appreciate and apply the ion chromatographic technique, it is necessary to have an understanding of the basics of the chromatography process. The representation of chromatographic signals in the form of a chromatogram generally has the appearance similar to that shown in Figure 2-1.

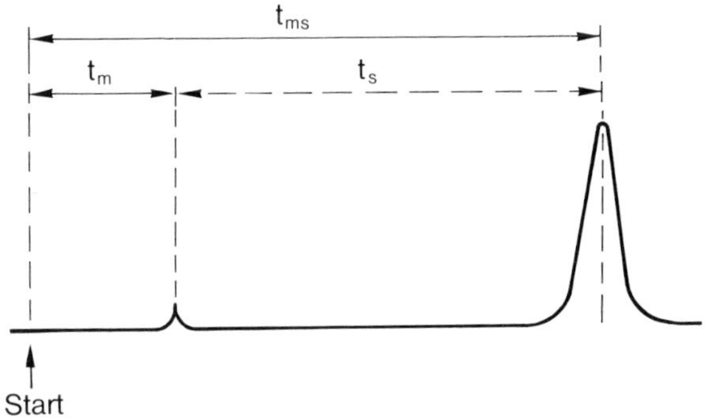

FIGURE 2-1. General illustration of a chromatogram.

The time from injection to elution of the maximum concentration of a compound which corresponds to the peak maximum, is denoted as the gross retention time, t_{ms}. The dead time, t_m, is the time required by a compound, which does not interact with the stationary phase, to migrate through the column. The net retention time, t_s, is given by the difference between the gross retention time, t_{ms}, and the dead time, t_m:

$$t_s = t_{ms} - t_m \tag{1}$$

The form of a chromatographic peak is usually approximated by a Gaussian curve (see Figure 2-2):

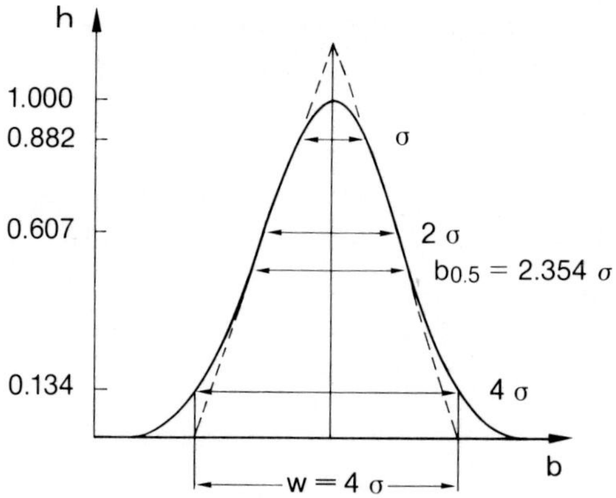

FIGURE 2-2.
The Gaussian curve.

σ: Standard deviation = half the peak width at the point of inflexion.
y = h: Peak height at maximum.
x: Any particular point within the peak.
μ: Position of the peak maximum.

The peak height at the point x can be calculated from Equation 2:

$$y = \left(\frac{1}{\sigma \cdot \sqrt{2\pi}}\right) \bullet \left(e^{\left(-\frac{(x-\mu)^2}{2\sigma^2}\right)}\right) \bullet (Y_o) \tag{2}$$

TAILING FACTOR, T

The signal due to the elution of a species from a chromatographic column is rarely perfectly Gaussian. Usually, the peak is asymmetric (see Figure 2-3) and this phenomenon is called "tailing."

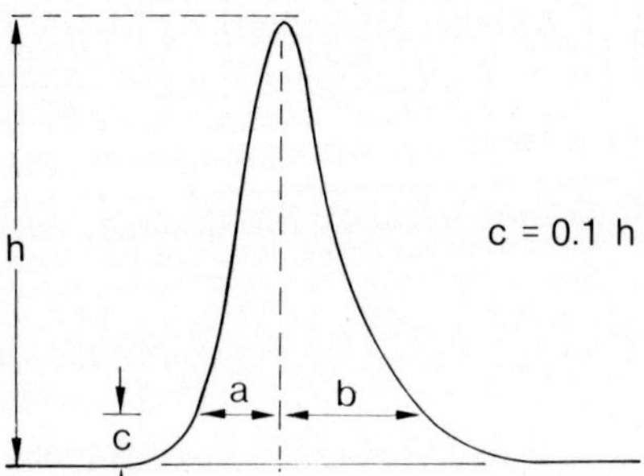

FIGURE 2-3.
Illustration of the tailing effect.

The degree of asymmetry—the tailing factor T—can be calculated from Equation 3:

$$T = \left| \frac{b}{a} \right| \tag{3}$$

The tailing effect, which can be ascribed mainly to adsorption processes, is characterized by a rapid increase of the detector signal and a comparatively slow decrease.

FRONTING

The opposite effect—a slow rise and a comparatively rapid decrease—is observed for the signals obtained by ion chromatography of alkaline earth metals and is termed "fronting." Fronting often occurs when some of the sample molecules (or ions) responsible for the signal cannot find free sorption sites and, thus, overtake the peak center.

Fronting can often indicate column overload.

2.2 PARAMETERS FOR ASSESSING THE QUALITY OF A SEPARATION

RESOLUTION

The goal of chromatography is to separate the sample components into separate bands, or peaks, as they migrate through the column. The resolution, R, between two adjacent peaks is defined by:

$$R = \frac{t_{ms_2} - t_{ms_1}}{\frac{w_1 + w_2}{2}} = \frac{2 \cdot \Delta t_{ms}}{w_1 + w_2} \tag{4}$$

$t_{ms_1}; t_{ms_2}$: Gross retention times for signals 1 and 2, respectively.

$w_1; w_2$: Peak width along the baseline as determined by the intersection points of the tangents drawn to the peak above its points of inflexion.

As shown in Figure 2-4, these parameters can be obtained directly from the chromatogram.

A resolution, R, equal to 0.5 is necessary for two components to be recognized as separate peaks. A complete separation of two peaks is obtained at $R > 1.5$. Further improvements in the resolution tend to lead to unnecessarily long analysis times.

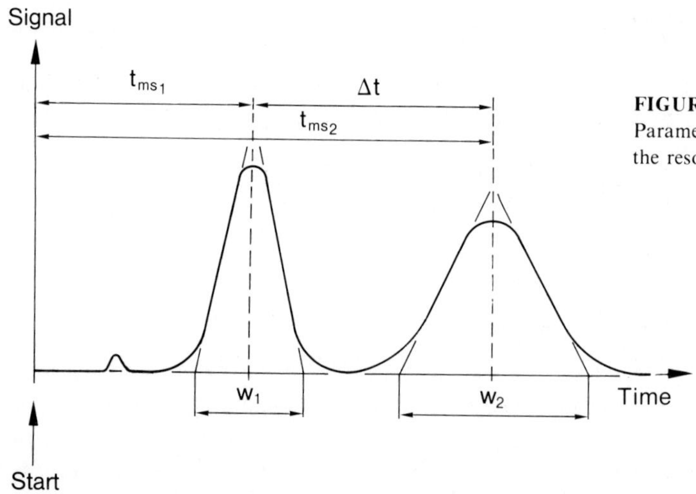

FIGURE 2-4. Parameters for assessing the resolution.

SELECTIVITY

The selectivity is given by the ratio of the net retention times of two signals and is a measure of their relative separation:

$$\alpha = \frac{t_{s_2}}{t_{s_1}} = \frac{t_{ms_2} - t_m}{t_{ms_1} - t_m} \tag{5}$$

This parameter can also be obtained from the chromatogram.

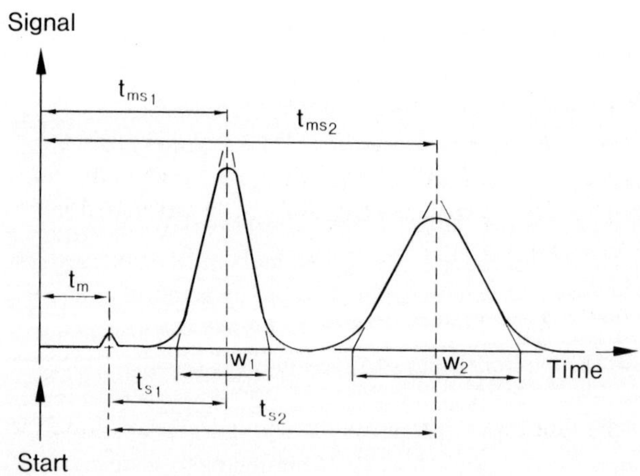

FIGURE 2-5. Parameters for assessing the selectivity.

CAPACITY FACTOR

The capacity factor, k′, is given by the product of the volume ratio of mobile and stationary phases in the separator and distribution coefficient, K:

$$k' = K \cdot \left(\frac{V_s}{V_m}\right) = \left(\frac{C_s}{C_m}\right) \cdot \left(\frac{V_m}{V_s}\right) = \left(\frac{t_{ms} - t_m}{t_m}\right) = \left(\frac{t_s}{t_m}\right) \tag{6}$$

K: Distribution coefficient.

V_s: Volume of the stationary phase.

V_m: Volume of the mobile phase.

C_s: Solute concentration in the stationary phase.

C_m: Solute concentration in the mobile phase.

Small values of k′ imply that the compound elutes near the void volume; thus, the separation will be very poor. Large k′ values indicate a good separation but also imply longer analysis times, an associated peak broadening, and a decrease in sensitivity.

2.3 COLUMN EFFICIENCY

By convention, a chromatographic column is divided into N theoretical plates. Within each plate, a thermodynamic equilibrium of the sample molecules between the mobile and the stationary phase is achieved. The plate height is related to the number of theoretical plates by Equation 7:

$$H = L/N \tag{7}$$

N: Number of theoretical plates.

L: Length of the column.

H is synonymous with HETP which stands for **H**eight **E**quivalent to a **T**heoretical **P**late. The value of H (defined as the ratio of the peak width and the gross retention time, t_{ms}) for a particular column can be calculated from a chromatogram as shown in Equation 8:

$$H = \frac{L}{8 \cdot \ln 2}\left(\frac{b_{0.5}}{t_{ms}}\right)^2 = \frac{L}{16}\left(\frac{w}{t_{ms}}\right)^2 \tag{8}$$

$b_{0.5}$: Peak width at half the peak height.

w: Distance between the intercepts of tangents drawn to the peak above the points of inflexion.

Combining Equations 7 and 8 leads to another expression for the number of theoretical plates:

$$N = \left(\frac{t_{ms}}{\sigma}\right)^2 = 16 \cdot \left(\frac{t_{ms}}{w}\right)^2 = 5.54 \cdot \left(\frac{t_{ms}}{b_{0.5}}\right)^2 \quad (9)$$

The factor 5.54 (8 ln 2) arises from the approximation of the chromatographic peak as a Gaussian curve. The resolution of a column can be calculated from the parameters—selectivity, capacity, and number of theoretical plates according to:

$$R = \underset{a}{\tfrac{1}{4}\sqrt{N}} \; \underset{b}{\left(\frac{\alpha - 1}{\alpha}\right)} \; \underset{c}{\left(\frac{k'}{k' + 1}\right)} \quad (10)$$

a: Term for the column efficiency.

b: Term for the column selectivity.

c: Term for the column capacity.

By rearrangement of Equation 10, the required number of theoretical plates for the desired resolution, R, can be calculated with given values for k′ and α:

$$N = 16\,R^2 \left(\frac{k' + 1}{k'}\right)^2 \left(\frac{\alpha}{\alpha - 1}\right)^2 \quad (11)$$

2.4 THE CONCEPT OF THEORETICAL PLATES (THE VAN-DEEMTER THEORY)

Due to increasing column-residence times, the peak broadening places an important limit on the chromatographic analysis. Based on the work of Martin and Synge [1], van-Deemter, Zuiderweg, and Klinkenberg [2] chose (in analogy to the terminology for distillation) the theoretical plate height HETP, H, as a measure of the relative peak broadening. A more general expression for HETP is given by Equation 12:

$$H = A + B\mu + C/\mu \quad (12)$$

μ: Linear velocity of the mobile phase in cm/s.

This equation emphasizes the different dependencies of the individual terms on the linear velocity of the mobile phase. According to Giddings [3], term A is independent on the flow rate and characterizes the peak broadening in terms of the multiple path effect. This effect considers the various pathways a species can take when migrating through the column and is dependent on the size and form of the packing material, as well as, on the quality of the column packing. The influence of the multiple path effect on the total peak broadening is given by Equation 13:

$$H_E = 2\lambda \cdot (d_p) \tag{13}$$

λ: A measure for the packing irregularity.

d_p: Particle diameter of the packing material.

Term B characterizes the lateral diffusion and the delay in the mass transfer processes which are often the dominating influences involved in total peak broadening. These processes represent the major determining factors for the efficiency of a chromatographic column. Each individual molecule, which interacts with the packing material, diffuses continuously towards the stationary phase and then back into the mobile phase. Due to the sorption of the molecule onto the stationary phase, it moves relatively slower through the column than the peak center. This effect is pictorially represented in Figure 2-6.

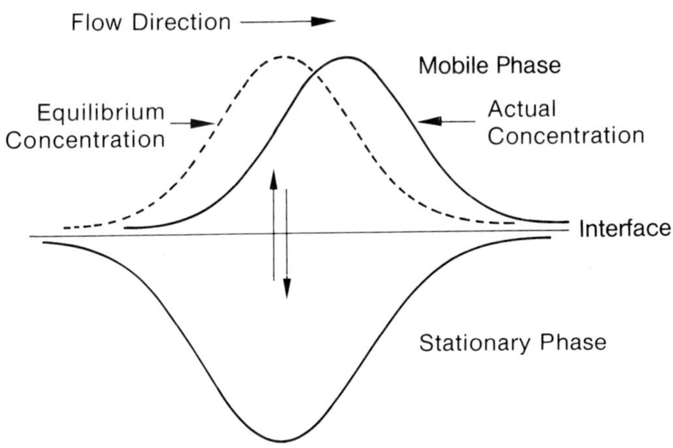

FIGURE 2-6. Illustration of the mass transfer effect.

In the mobile phase, the solute is carried along with the speed of the eluant. The mass transfer effect leads to peak broadening since some of the sample molecules migrate through the column in front of, and some behind, the peak maximum. Due to the eluant flow within the column, the concentration of the solute at any point is not in equilibrium with that at the surface of the stationary phase. Thus, both phases are responsible for the retardation of the mass transfer and two terms are necessary to describe the peak broadening. The broadening due to the stationary phase is given by Equation 14:

$$H_s = \frac{f \cdot p \cdot (1 - p) \cdot d_f^2}{D_s} \cdot \mu \qquad (14)$$

f: Form factor for the stationary phase.
p: Solute concentration in the stationary phase.
$(1 - p)$: Solute concentration in the mobile phase.
d_f: Thickness of the stationary phase.
D_s: Diffusion coefficient of the solute in the stationary phase.

Pellicular materials, such as are used in IC, minimize the contribution of the stationary phase to the total peak broadening because the solute molecules cannot penetrate into the inner region of the stationary phase. Thus, an apparent quadratic relationship between the peak broadening, H_s, and the thickness of the stationary phase, d_f, can be explained.

The dependence of the peak broadening on the mobile phase is given by Equation 15:

$$H_M = \frac{\omega \cdot d_p^2}{D_m} \cdot \mu \qquad (15)$$

d_p: Particle diameter.
D_m: Diffusion coefficient of the solute in the mobile phase.
ω: Measure of the quality of the packing process.

The contribution of H_M to the total peak broadening is important; but by reducing the particle diameter, d_p, this contribution can be significantly reduced. Using an eluant with a lower viscosity also reduces the peak broadening, H_M, due to a corresponding increase in the diffusion coefficient of the solute molecules. Larger values of D_m can also be achieved by increasing the column temperature, but the associated increase in column efficiency tends to lead to a reduced selectivity.

The longitudinal diffusion of the sample molecules in the mobile phase also contributes to the broadening of the peaks and is described by term C:

$$H_L = 2\gamma\,(D_m) \cdot (1/\mu) \qquad (16)$$

γ: Obstruction factor.
D_m: Diffusion coefficient of the solute in the mobile phase.

The diffusion of the solute molecules in, and against, the direction of flow of the mobile phase makes a negligible contribution to the total peak broadening in liquid chromatography.

By combining Equations 13–16, Equation 17 defining the effective HETP is obtained:

$$H = H_E + H_S + H_M + H_L$$
$$= 2\lambda \cdot d_p + \frac{f \cdot p \cdot (1-p) \cdot d_f^2}{D_s} \cdot \mu + \frac{w \cdot d_p^2}{D_m} \cdot \mu + 2\gamma\, D_m\, (1/\mu) \quad (17)$$

2.5 VAN-DEEMTER CURVES IN ION CHROMATOGRAPHY

The combined terms in Equation 17, denoting the influence of the various factors on the total peak broadening, give the impression that they are independent of one another. However, in practice, it appears that a certain dependence does exist. Thus, the decrease in the performance of the column with greater flow rates is considerably less than that calculated on the basis of the simplified van-Deemter Theory.

A plot of the calculated values for H (from Equations 7 and 9) versus the flow rate, μ, for an anion exchange separator, HPIC-AS4, is shown in Figure 2-7.

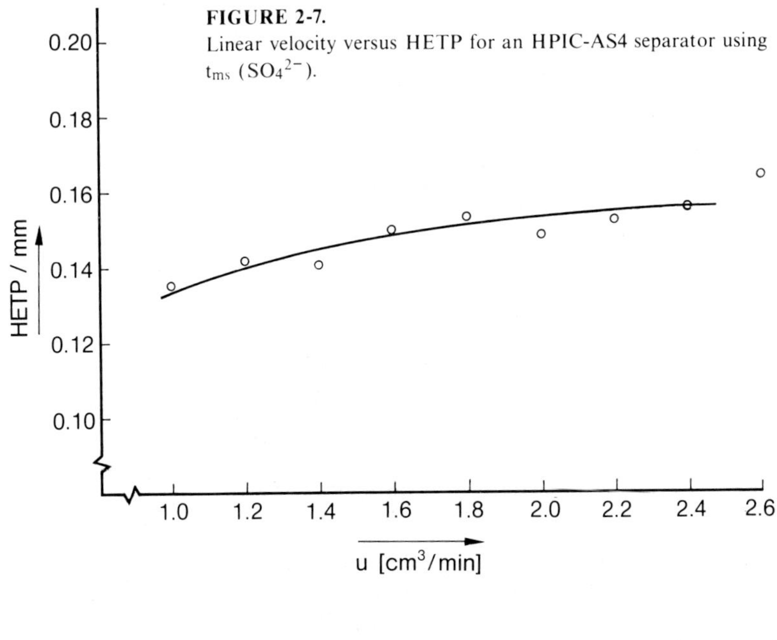

FIGURE 2-7.
Linear velocity versus HETP for an HPIC-AS4 separator using t_{ms} (SO_4^{2-}).

From Figure 2-7, it is clear that, at greater flow rates, HETP remains essentially constant. A similar observation has been made by Majors [5] for HPLC columns, based on silica with small particle sizes. Such relationships show that it is possible to save analysis time by increasing flow rate without a significant loss in resolution.

For cation exchange columns of the type HPIC-CS1, packed with larger particles, a pronounced dependence of HETP on the flow rate is observed (see Figure 2-8).

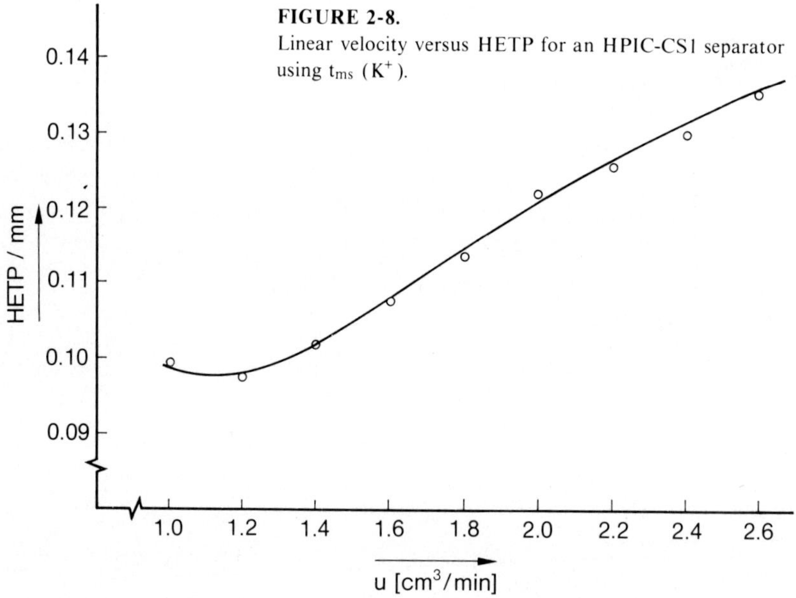

FIGURE 2-8.
Linear velocity versus HETP for an HPIC-CS1 separator using t_{ms} (K^+).

In this case, a compromise must be made between the resolution and the time required for analysis. Flow rates between 2.0 and 2.3 cm^3/min have generally proved to be optimal.

LITERATURE

1. A.J.P. Martin and R.L.M. Synge, Biochem. J. **35,** 1358 (1941).
2. J.J. van Deemter, F.J. Zuiderweg and A. Klinkenberg, Chem. Eng. Sci. **5,** 271 (1956).
3. J.C. Giddings, "Dynamics of Chromatography," Part 1, Marcel Dekker, New York (1965).
4. E. Johnson and R. Stevenson, "Basic Liquid Chromatography," Varian Associates, Inc., Palo Alto, Ca. (1978).
5. R.E. Majors, Anal. Chem. **44,** 1722 (1972).

3 Ion Exchange Chromatography (HPIC)

Ion exchange is one of the oldest separation processes described in the literature [1,2]. It is derived from classical column chromatography using microporous ion exchange resins. Modern ion exchange chromatography differs from classical column chromatography in the way in which the sample is introduced onto the column, the material being used, and the detection system. Following classical techniques a column, some 10–50 cm long, was packed with either an anion or cation exchange resin with particle diameters of 250-75 microns. After the sample (often several milliliters) had been applied to the top of the column, it migrated through the column with the eluant flow under the influence of gravity and became more or less separated. The individual fractions were collected, using an automatic sample collector, and subsequently analyzed, using wet chemical techniques. Due to the high ion exchange capacity of those resins, high eluant concentrations were necessary to ensure the elution of the ions. The amount of eluant required for analysis was often several liters.

The enormous improvement in the capability of modern ion exchange chromatography can be attributed to the pioneering work of Small and coworkers [3] who developed techniques for the reproducible preparation of ion exchange resins with low capacities and *high efficiencies*. The required sample volume was reduced to 10–100 μl, allowing improved resolution due to very narrow peaks. One of the most important developments was that of automated detection, which allowed the continuous monitoring of the eluate. The introduction of *conductivity detectors* for the detection of ionic species provided new dimensions to ion chromatography.

3.1 THE ION EXCHANGE PROCESS

The resins used in ion chromatography are functionalized with a fixed charge and an associated counterion which makes the resin electrically neutral. The exchange function in anion chromatography is generally a quarternary ammonium base. In cation chromatography, it is a sulphonate ion. During the chromatographic process, the counterion of the exchange function is exchanged for the sample ion and, for a short time, the latter becomes attached to the fixed charge on the resin. Due to variations in the affinities of the species being chromatographed for the stationary phase, a separation of the various sample components is possible.

For example, if an anion exchange resin is treated with a solution containing the anion HCO_3^-, the fixed $-\overset{+}{N}R_3$ groups are converted completely into the hydrogen carbonate form. If a solution containing the anions A^- and B^- is now injected onto the separator, these anions will be exchanged in an equilibrium process for the hydrogen carbonate anions (see Equations 18 and 19):

$$\text{resin-}\overset{+}{N}R_3\ HCO_3^- + A^- \underset{}{\overset{K_1}{\rightleftharpoons}} \text{resin-}\overset{+}{N}R_3\ A^- + HCO_3^- \tag{18}$$

$$\text{resin-}\overset{+}{N}R_3\ HCO_3^- + B^- \underset{}{\overset{K_2}{\rightleftharpoons}} \text{resin-}\overset{+}{N}R_3\ B^- + HCO_3^- \tag{19}$$

The separation of the anions will be controlled by their different affinities for the stationary phase. The selectivity coefficient which characterizes the exchange process can be defined by:

$$K = \frac{[X^-]_s\,[HCO_3^-]_m}{[HCO_3^-]_s\,[X^-]_m} \tag{20}$$

$[X^-]_{m,s}$: Concentration of the sample ion in the mobile (m) or the stationary phase (s).

$[HCO_3^-]_{m,s}$: Concentration of hydrogen carbonate in the mobile (m) or the stationary phase (s).

Experimentally, the selectivity coefficient can be determined by equilibrating a known amount of a solution containing the ions X^- and HCO_3^- with a known weight of the ion exchange resin. After allowing equilibrium to be attained, the concentration of the ions in the mobile and stationary phases can be calculated from Equation 20. For any ion X^- a weight distribution coefficient (D) can be defined according to Equation 21:

$$D = \frac{[X^-]_s}{[X^-]_m} \tag{21}$$

Rather than the weight distribution coefficient, it is more common to use the capacity factor k′, which is related to D as shown in Equation 22:

$$k' = D\left(\frac{\text{g resin}}{\text{cm}^3\ \text{solution}}\right) \tag{22}$$

k′ is usually determined from a chromatogram as described before (see Equation 6 in Chapter 2, Section 2.)

3.2 ANION EXCHANGE CHROMATOGRAPHY

THE STRUCTURE OF THE STATIONARY PHASE

The Structure of the Support Material

Stationary phases, used in HPLC, are usually based on silica; whereas, in ion chromatography, the support material is a polystyrene/divinylbenzene based resin which is much more stable with respect to pH than silica. Thus, whereas silica can be used between pH 3 and 9, IC separators are stable between pH 0 and 14. The pH stability allows extreme pH values to be used that can convert molecular compounds—such as sugars and alcohols—into ionic species. The copolymerization of styrene with DVB is necessary to give the resin a mechanical stability. The amount of DVB in the resin is denoted as "percent crosslinking."

The degree of crosslinking is directly related to the extent to which PS/DVB resins shrink or swell in aqueous media, or especially in the presence of organic solvents such as acetonitrile. Variation of the eluant composition can also lead to swelling and shrinking problems. If the resin contracts, it becomes more tightly packed; leading to a dead volume at the beginning of the column which, in turn, results in a drastic loss in column efficiency. In contrast, swelling of the resin leads to higher column backpressures when constant flow rates are to be maintained. As a general rule, a resin with a higher degree of crosslinking swells less. Experimentally, it has been shown that the optimum degree of crosslinking is between 2 and 5%. Higher degrees of crosslinking are disadvantageous, because the porosity of the resin decreases with an increasing amount of DVB. This, in turn, limits the mass transfer and, consequently, leads to a loss in column efficiency.

The Structure of Anion Exchange Resins

The structure of the anion exchange resins used by Dionex and developed by Small is illustrated in Figure 3-1.

These anion exchange resins are composed of a surface sulphonated PS/DVB core with a particle diameter between 10 and 25 μm, and a totally porous "latex" particle which is completely aminated. The latex particles have a considerably smaller diameter of about 0.1 μm and carry the actual ion exchange function $-\overset{+}{N}R_3$. The agglomeration of latex particles onto the core employs both electrostatic and van-der-Waals interactions. Thus, the ion exchange resin is composed of three, chemically very different, regions:

- An inert, mechanically stable core.
- A film of sulphonic acid groups covering the core completely.
- An outer layer of latex particles carrying the exchange functions.

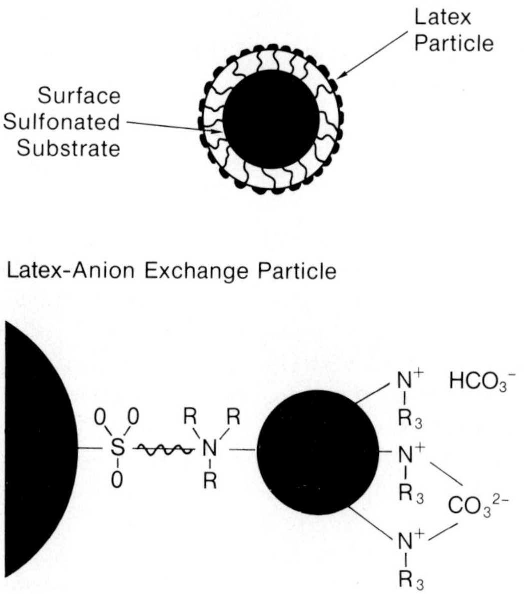

FIGURE 3-1.
The structure of a latex anion exchange resin.

According to Horvath [4], such packing material can be denoted as a superficial ion exchange resin in which the diffusion paths are determined by two factors:

- The size of the latex particles.
- The degree of surface sulphonation.

The particle size of the latex material determines the length of the diffusion paths and, thus, the rate of the diffusion process. Higher degrees of surface sulphonation tend to hinder the diffusion of other anionic species into the stationary phase due to Donnan exclusion (see Chapter 4) so that the functional groups bound to the latex particle dominate the diffusion process. This complex system has a number of advantages over other column materials, such as silica-based anion exchange resins (known under the trade name "Vydac"), and directly aminated PS/DVB resins developed by Fritz and coworkers [5].

Thus:

- The inner core guarantees the mechanical stability of the resin and ensures moderate column backpressures.
- Rapid exchange processes and, hence, high efficiencies are ensured by the small, totally porous, latex particles.
- Swelling and shrinking of the material is greatly reduced due to the surface sulphonation.

Latexed anion exchange resins are distinguished by an enormous chemical stability. Even 4*M* NaOH does not break the bond between the latex and the core. The selectivity of the resins can be varied by using different quarternary ammonium bases. Since the latex material is prepared separately, the selectivity of a column can easily be optimized for a particular analytical problem by varying the quarternary ammonium base.

For the user, these materials have the following advantages: They are easy to handle; they can be used with inexpensive eluants; and, in particular, a wide range of selectivities is available.

Ion Exchange Capacity: The ion exchange capacity of a resin is defined as the number of exchangeable groups per unit weight of the resin and is generally quoted in milliequivalents/g resin (mequiv./g). It is generally true that the greater the exchange capacity of the resin, the longer it takes for the sample ions to elute. This effect can be partly compensated by using higher eluant concentrations.

REVIEW OF THE VARIOUS ANION EXCHANGE RESINS AVAILABLE

At the time of this writing, seven different anion exchange resins, with various selectivities, are available from Dionex (HPIC-AS1 to AS7). The characteristic structural and technical properties of these materials are summarized in Table 3-1.

TABLE 3-1: Comparison of the various anion exchange separators available from Dionex.

COLUMN	PARTICLE SIZE OF THE PACKING MATERIAL	DEGREE OF CROSSLINKING	SIZE OF THE LATEX PARTICLES	HYDROPHOBICITY OF THE FUNCTIONAL GROUPS
HPIC-AS1	25 μm	5%	m	−
HPIC-AS2	25 μm	5%	m	+
HPIC-AS3	25 μm	2%	s	+
HPIC-AS4	15 μm	3.5%	s	+
HPIC-AS5	15 μm	1%	s	−
HPIC-AS6	10 μm	5%	l	+
HPIC-AS7	10 μm	5%	l	+

s = small, m = medium, l = large.
\+ = hydrophobic, − = hydrophilic.

The different selectivities of the columns are particularly influenced by the following key parameters:

- Degree of crosslinking of the latex.
- Size of the latex particles.
- Type of latex functionality.

Figure 3-2 shows how a separation of mineral acids is affected by changing these parameters.

Although the only difference between the separators AS1 and AS2 is the hydrophobicity of the latex, a different elution sequence is observed for bromide, nitrate, and sulphate ions. Both bromide and nitrate are retained for some time on the stationary phase, but their separation and elution sequence only depends on the different degrees to which they are adsorbed onto the stationary phase. This effect is superimposed on their ion exchange characteristics. Changing the hydrophobicity of the functional groups thus affects the retention behavior of both these ions selectively. Such adsorption effects can be demonstrated by a simple experiment: If the adsorption sites on the stationary phase are blocked by adding p-cyanophenol—which is particularly suitable for this purpose—to a bicarbonate/carbonate eluant, then bromide and nitrate coelute as is shown in Figure 3-3.

If, additionally, the size of the latex particles and the degree of crosslinking are reduced, the elution profile of an AS3 separator is obtained. A further reduction

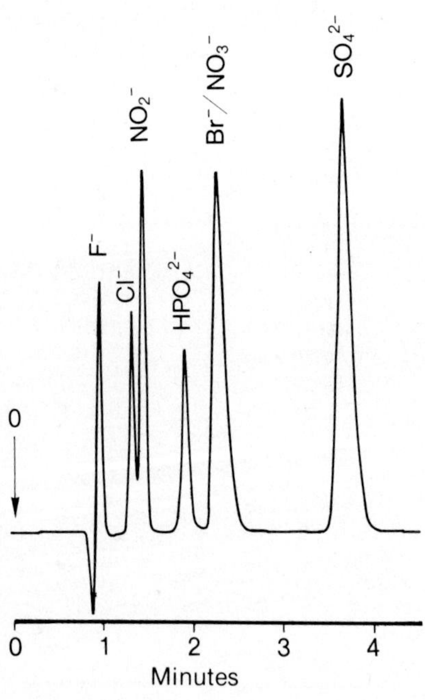

FIGURE 3-3.
Chromatogram showing the effects of adsorption on a separation of bromide and nitrate. **Eluant:** 0.0043 mol/dm^3 $NaHCO_3$; 0.0034 mol/dm^3 Na_2CO_3 and 100 mg/dm^3 p-cyanophenol. **Flow rate:** 2 cm^3/min. **Detection:** Suppressed conductivity. **Sample:** 50 μl anion standard.

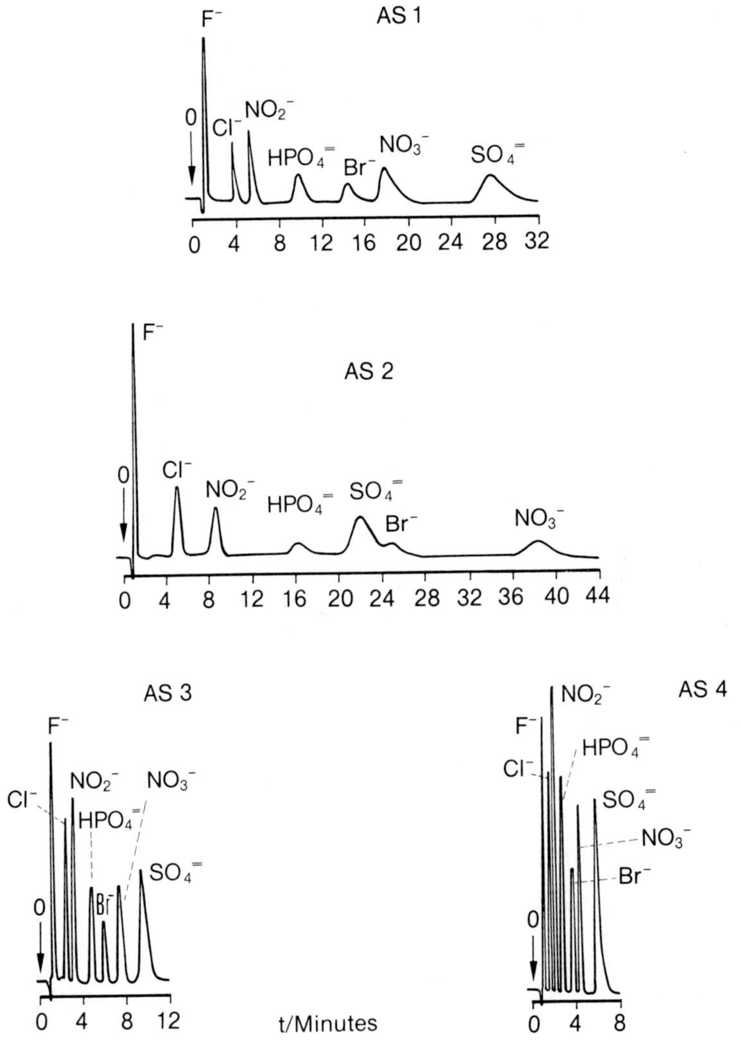

FIGURE 3-2.
A comparison of the selectivity of the HPIC-AS1 to AS4 anion exchange columns. **Eluant:** 0.0028 mol/dm^3 $NaHCO_3$ and 0.0022 mol/dm^3 Na_2CO_3. **Flow rate:** 2 cm^3/min. **Detection:** Suppressed conductivity. **Sample volume:** 50 μl. **Anion standard:** 3 ppm F^-, 4 ppm Cl^-, 10 ppm NO_2^-, 25 ppm HPO_4^{2-}, 10 ppm Br^-, 20 ppm NO_3^- and 25 ppm SO_4^{2-}.

of the retention times can be achieved by increasing the column efficiency using a core material with a smaller particle size. But this, in turn, leads to a considerable increase in the column backpressure. Comparing the performance of an AS1 and AS4 separator, the time required for a baseline resolved separation of the seven different anions is reduced by ca. 60% in this way.

Increasing the efficiency of an ion exchange separator also leads to a loss in exchange capacity so that there is little point in further increasing the efficiency by using even smaller particles. This would mean that even a simple drinking water sample would have to be diluted some 100 fold in order to avoid an overloading of the column. The time required for such a sample preparation, and the possible statistical errors associated with dilutions, would not justify the shortening of the analysis time.

For the analysis of hydrophobic anions—such as, iodide, thiocyanate, thiosulphate, etc.—the AS5 separator was developed. Figure 3-4 shows a comparison of the selectivities of an AS4 and AS5 column for the separation of mineral acids.

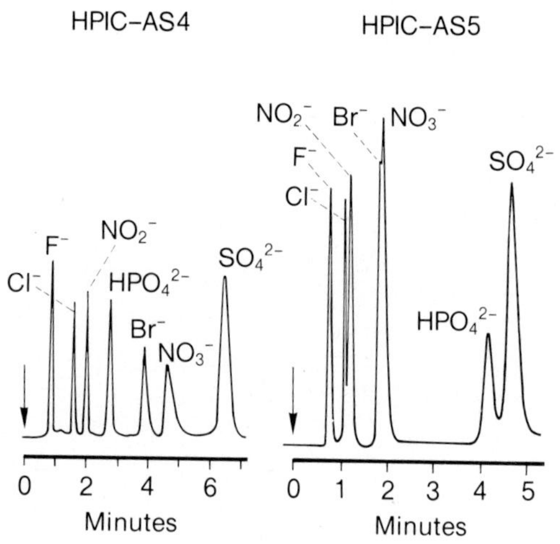

FIGURE 3-4.
A comparison of the selectivity of columns HPIC-AS4 and AS5. **Eluant:** 0.0028 mol/dm^3 $NaHCO_3$ and 0.0022 mol/dm^3 Na_2CO_3. **Detection:** Suppressed conductivity. **Sample:** 50 μl anion standard.

Since the particle size in both columns is the same, the resulting analysis times are roughly the same. The difference in the selectivities is due to the hydrophilicity of the functional groups which makes a separation of bromide and nitrate impossible. Hydrophilic functional groups are, however, a prerequisite for the separation of *hydrophobic* anions, which, because of their large radii, are strongly adsorbed onto the stationary phase. Therefore, the resulting peaks tail badly. The separation of iodide and thiocyanate ions, shown in Figure 3-5, clearly shows that the tailing effect can be suppressed by adding p-cyanophenol.

Carbohydrates and polyvalent anions can be analyzed utilizing the separators AS6 and AS7, respectively. These columns have only recently been developed and contain, for the first time, a 10 μm core material. Both packing materials are characterized by relatively low exchange capacities and large latex particles carrying hydrophobic functional groups.

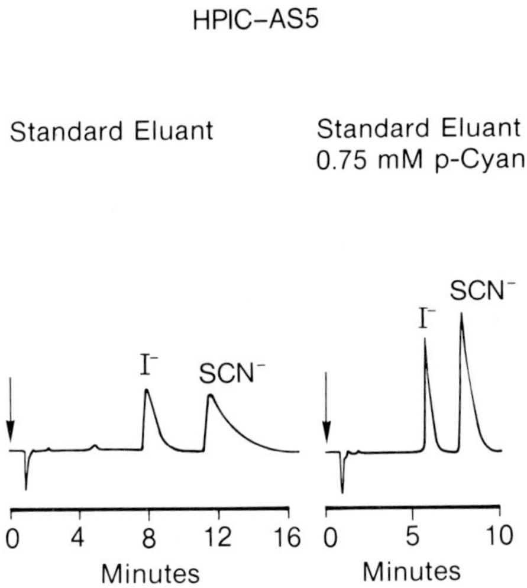

FIGURE 3-5.
The influence of p-cyanophenol on iodide and thiocyanate peak tailing. The chromatographic conditions used were the same as those used to obtain the chromatogram shown in Figure 3-3.

The analysis of carbohydrates by ion exchange chromatography requires that they are first converted into their ionic forms. Since the pK values of carbohydrates lie between 13 and 14, 0.10–0.15M NaOH concentration is used as eluant. Compared with the chemically bonded amino phase, which is used in HPLC, the elution sequence of carbohydrates in ion chromatography is reversed. Sugar alcohols elute before mono-, di-, and oligosaccharides.

Polyvalent anions—such as NTA, EDTA, pyrophosphate, triphosphate and polyphosphonates—previously could not be analyzed with ion chromatography because the retention time of an ion increases with increasing valency. Furthermore, large adsorption effects result in unacceptable tailing effects. By combining a suitable functionality with an appropriate latex particle size and employing a mobile phase of a high elution strength, it is possible to obtain totally symmetrical peaks for polyvalent anions using an AS7 separator.

ELUANTS FOR ANION CHROMATOGRAPHY

The eluants suitable for anion chromatography can be divided into two basic categories:

- Eluants suitable for nonsuppressed conductivity detection.
- Eluants suitable for suppressed conductivity detection.

Both categories have the following properties in common:

- The affinity of the eluant and the sample ion for the stationary phase should be compared.
- The product of the suppressor reaction must be a species with a low conductivity. Or: The eluant itself must have a low conductivity in order to allow a sensitive conductivity detection of the sample anions.

In this chapter, only those eluants which are compatible with suppressor systems will be discussed. The other category will be discussed in the section "Related Ion Chromatographic Techniques."

For his fundamental work in 1975, Small [3] used sodium hydroxide and/or sodium phenate solutions as eluant for the analysis of hydrophilic, inorganic anions. In the meantime, they have been replaced by the more versatile bicarbonate/carbonate system. Sodium hydroxide has the disadvantage that it exhibits a low affinity for the stationary phase and, thus, relatively large concentrations which imply high background conductivities, have to be used.

Using sodium phenate as the eluant does give good separations for inorganic anions, but even traces of oxygen in the mobile phase lead to oxidation of the sodium phenate and poisoning of the packing material by the oxidation products. An additional disadvantage of this system is the toxicity of phenol as the product of the suppressor reaction.

In contrast, sodium bicarbonate/carbonate mixtures have proved particularly useful because the strength of the eluant—and thus the selectivity of the system—can be adjusted simply by varying the proportion of the two compounds. Furthermore, the product of the suppressor reaction is H_2CO_3 which is only weakly dissociated and, thus, the resulting background conductivity is very low.

Analysis of aliphatic fatty acids with a small number of C-atoms, which elute close to the dead volume, requires eluants of a low elution strength such as sodium tetraborate or sodium bicarbonate. A survey of the various eluant compositions for the analysis of hydrophilic anions and their elution strengths is given in Table 3-2.

TABLE 3-2: Eluants for anion chromatography.

ELUANT	ELUTING ION	SUPPRESSOR PRODUCT	ELUTION STRENGTH
$Na_2B_4O_7$	$B_4O_7^{2-}$	H_3BO_3	Very Weak
NaOH	OH^-	H_2O	Weak
$NaHCO_3$	HCO_3^-	H_2CO_3	Weak
$NaHCO_3/Na_2CO_3$	HCO_3^-/CO_3^{2-}	H_2CO_3	Medium
Na_2CO_3	CO_3^{2-}	H_2CO_3	Strong
$NaNO_3$	NO_3^-	—	Very strong
NaI	I^-	AgI	Very strong

Hydrophobic anions are also separated using a bicarbonate/carbonate mixture as the eluant; but for such ions, p-cyanophenol should be added to the mixture.

The analysis of polyvalent anions can be achieved using a relatively concentrated nitric acid solution.

In general, reagent grade chemicals usually provide good results; but when nitric acid is used, an ultrapure quality is recommended since polyphosphates and polyphosphonates are complexing agents and interact with heavy metals still present in reagent grade nitric acid. Such interactions would be detrimental to the analytical results.

SUPPRESSOR SYSTEMS IN ANION CHROMATOGRAPHY

In order to obtain a sensitive detection of ions via their conductivity, Small used a suppressor column that can be considered part of the detection system. The function of the suppressor column can be explained in a better way by looking at the example of the separation of chloride and bromide anions. After separation via anion exchange, such as has already been described, the column effluent is directed into a second column which contains a strongly acidic cation exchange resin in the hydrogen form. In this second column the following reactions occur:

1. Exchange of sodium ions for protons converts the highly conductive sodium bicarbonate into the weakly dissociated carbonic acid. This reaction is denoted as suppression of the eluant conductivity.

$$\text{resin-}SO_3^-H^+ + NaHCO_3 \rightarrow \text{resin-}SO_3^-Na^+ + (CO_2 + H_2O) \qquad (23)$$

2. Sodium chloride and sodium bromide are converted into the corresponding acids:

$$\text{resin-}SO_3^-H^+ + NaCl \rightarrow \text{resin-}SO_3^-Na^+ + HCl \qquad (24)$$

$$\text{resin-}SO_3^-H^+ + NaBr \rightarrow \text{resin-}SO_3^-Na^+ + HBr \qquad (25)$$

Thus, as the result of the suppressor reaction, the highly conductive inorganic mineral acids are detected in the presence of only weakly conductive carbonic acid with considerable sensitivity.

An additional advantage of the suppressor technique, compared with those techniques which do not employ such a unit, is that the former have a greater specificity. Thus, Jenke and Pagenkopf [6] have shown that heavy metals—such as lead, zinc, and copper—coelute with inorganic anions such as chloride, bromide, and nitrate. Under particular conditions, this can lead to false interpretations of the chromatographic results.

The quality of the separation is influenced by the following properties of the suppressor column:

- The *pore volume* is critical for the analysis of weak electrolytes, such as nitrite or acetate, which are not affected by Donnan exclusion. (The pore volume is determined by the degree of crosslinking of the resin and the total volume of the suppressor column.)
- The *total surface area of the resin* in the suppresor column which is responsible for adsorption processes.
- The *particle diameter of the resin* and *the dead volume* of the column which determine the degree of peak broadening.

To some extent these parameters are mutually dependent and, therefore, they must be optimized in their relationship to one another. (For example, a higher degree of crosslinking minimizes the pore volume, but leads to increased adsorption effects.) A more important disadvantage of the conventional suppressor columns is that they require a periodic regeneration. Thus, the retention time of the negative water peak depends on the extent to which the suppressor is expended, and considerable difficulties arise with the quantification of the fluoride and chloride signals. The Donnan effect, which has already been mentioned, has a negative effect on the routine analysis of weak electrolytes; since a fully regenerated cation exchange resin, in the hydrogen form, resembles an ICE column. Hence, weak acids, which are only partly dissociated, can penetrate the Donnan membrane and interact with the stationary phase. As the degree of expension of the suppressor column increases, the exchange groups are more and more converted into the sodium form which is not capable of adsorbing undissociated acids. This phenomenon leads to variation in peak heights.

Due to the development of hollow fiber suppressors, these disadvantages have been eliminated. The fiber suppressor, depicted in Figure 3-6, consists of a cylindrically-shaped body around which a semipermeable fiber membrane is wrapped.

The column effluent is directed through the interior of the fiber that is packed with inert beads. A dilute acid regenerent solution flows countercurrently to the eluant, in contact with the exterior of the fiber. The flow is gravity fed into the fiber; the flow rate controlled by the difference in height between the reservoir and the suppressor housing. When working in extremely sensitive detector ranges (1 μS/cm to 0.3 μS/cm full scale), it is recommended to pump the regenerent solution through the fiber suppressor, since the constantly changing hydrostatic pressure leads to baseline drifts. This type of suppressor does not require periodic regeneration. The reaction taking place within the semi-permeable membrane of a hollow fiber suppressor is also shown in Figure 3-6.

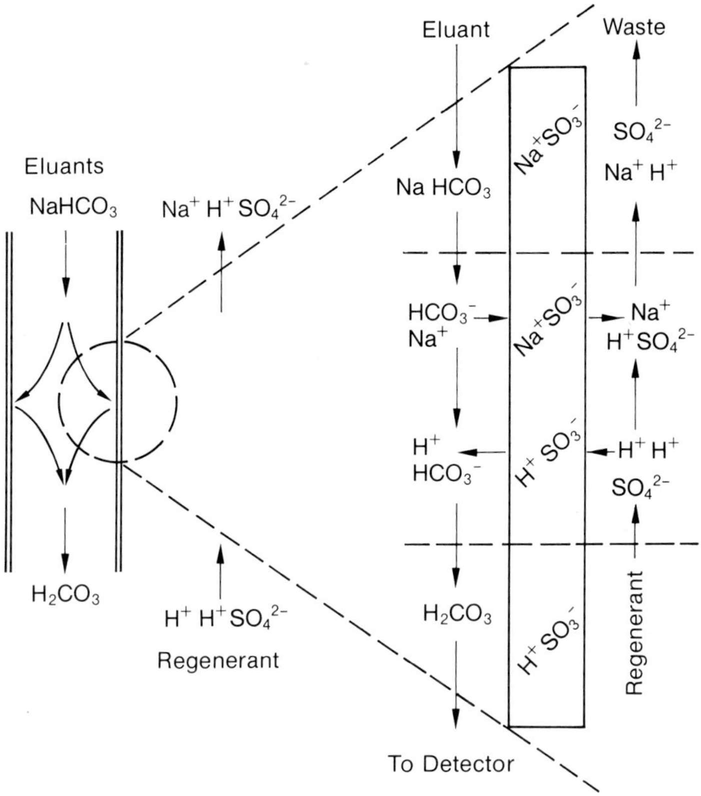

FIGURE 3-6.
The way an anion membrane suppressor works.

Since an anion suppressor is a cation exchanger, the cations in the eluate are exchanged for protons in the regenerent solution. The driving force for the diffusion of protons across the membrane is provided by the subsequent reaction of bicarbonate to carbonic acid. In order to maintain an ionic balance, the cations diffuse into the regenerent solution. Suppression is, therefore, maintained.

As shown in Figure 3-6, these hollow fibers can be divided into three different regions.

- An expended region where the eluate enters the fiber suppressor (cation exchanger in the sodium form).
- A region of dynamic equilibrium in which the actual suppression reactions occur.
- A regenerated region where the eluate exits the fiber suppressor (cation exchanger in the hydrogen form).

Once established, the dynamic equilibrium will be maintained as long as pertinent operating conditions are not altered. Since the percentage of the exchange groups in the sodium form remains virtually constant, weak acids also remain constant in response and less standardization is required. Additionally, the quantification of fluoride and chloride is simplified, since the water dip immediately precedes the fluoride peak. Because of the low dead volume of a membrane suppressor, mixing and band broadening effects are minimized and the sensitivity is generally enhanced compared to the more traditional packed bed suppressor as shown in Figure 3-7.

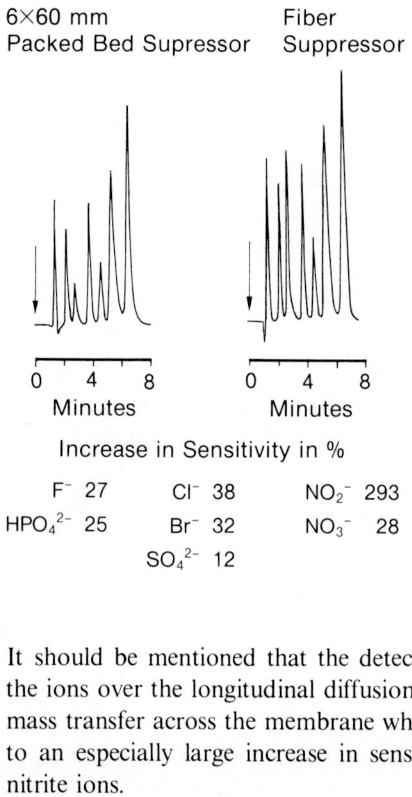

FIGURE 3-7.
A comparison of the limits of detection using a conventional suppressor column and a membrane suppressor. **Column:** HPIC-AS4. **Eluant:** 0.0028 mol/dm^3 $NaHCO_3$ and 0.0022 mol/dm^3 Na_2CO_3. **Flow rate:** 2 cm^3/min. **Detection:** Conductivity with (a) suppressor column ASC-1 and (b) membrane suppressor AFS-1 with a 2 cm^3/min flow of 0.0125 mol/dm^3 H_2SO_4 as regenerating solution. **Sample:** 50 μl anion standard.

It should be mentioned that the detection favors the translational diffusion of the ions over the longitudinal diffusion (Figure 3-8). This, in turn, improves the mass transfer across the membrane which, as can be seen from Figure 3-9, leads to an especially large increase in sensitivity with respect to the weakly acidic nitrite ions.

Typical Operating Conditions

Conventional packed bed suppressors of the type ASC-1 and ASC-2 are normally regenerated with 0.5M sulphuric acid at a flow rate of 4 ml/min for about 15 minutes. After regeneration, the suppressor columns are rinsed with deionized water for another 15 minutes. For the continuous regeneration of a fiber suppressor, 0.0125M sulphuric acid, at a flow rate of between 2 and 3 ml/min, is recommended. With these standard conditions, a background conductivity of between 15 and 25 μS/cm can be achieved using a bicarbonate/carbonate eluant (0.0028M and 0.0022M, respectively). The concentration of the

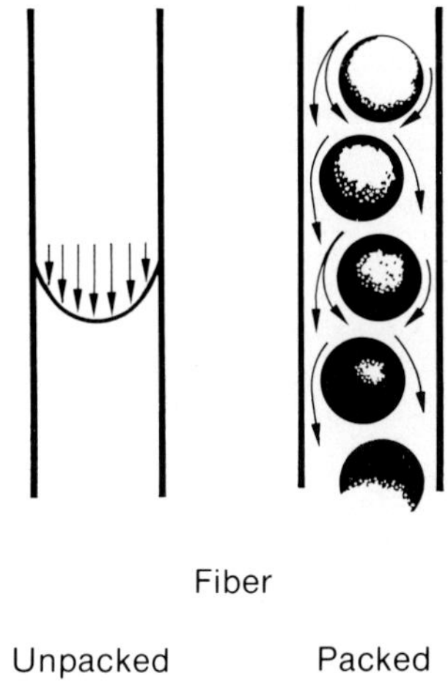

FIGURE 3-8.
Scheme of the flow-profile across a membrane suppressor.

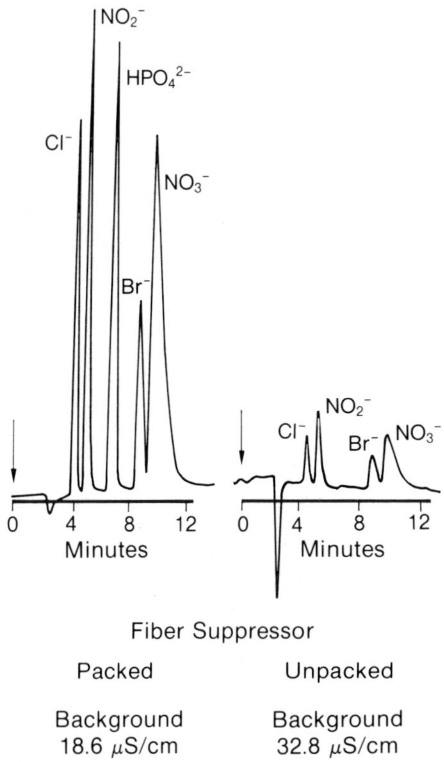

FIGURE 3-9.
A comparison of the detection limits using a membrane suppressor with and without packing.

regenerent solution should not be altered when working with less concentrated eluants. This will result in a correspondingly lower background conductivity.

When using very strong eluants, it is possible to totally expend the fiber suppressor. However, it is not possible to increase the regenerent concentration very much since, at higher concentrations, the sulfate ions can overcome the Donnan exclusion forces and diffuse into the fiber. Table 3-3 lists the background conductivities which result from regenerent solutions of different concentrations.

TABLE 3-3: Dependence of the background concentration on the regenerant concentration.

H_2SO_4 (mol/dm^3)	BACKGROUND CONDUCTIVITY
0.0125	21.7 μS/cm
0.0250	28.2 μS/cm
0.1250	288 μS/cm

Although higher concentrations of Na_2CO_3 can be suppressed, the limiting factor will be the inability to add an increased backpressure to prevent CO_2 outgassing inside the conductivity cell. Because of the limited pressure stability of the fiber, the backpressure should not exceed ca. 200 psi (13 bar).

When working with extremely strong eluants such as Na_2CO_3 ($C > 0.008M$), it is advantageous to use dodecyl benzene sulphonic acid (DDBSA) as regenerent. For example, when using sodium carbonate solutions with a concentration of 0.01M at a flow rate of 2 ml/min, a 0.1M DDBSA regenerent solution is recommended. With a regenerent flow rate of 3 ml/min, a background conductivity of some 50 μS/cm results. Disadvantages of DDBSA are the excessive surface activity (→ foam building) and the relative impurity of commercially available products (84–88%).

ION EXCHANGE CHROMATOGRAPHY OF INORGANIC ANIONS

Hydrophilic Anions

Terminology

Compared with other liquids, water is a particularly good solvent for salts. The reason for that is the special structure of water and the way in which the water molecule can interact with the individual ions. As salt is dissolved in water, hydrogen bonds are broken and the water loses some of its structure (cavity effect). The larger the ion, the greater the amount of energy required for the formation of the appropriate cavity. On the other hand, a new structure is formed due to ion dipole interactions. These effects are greater for ions having smaller radii and greater charge densities. As can be seen from the values of the molar enthalpy of hydration (see Table 3-4), the relative hydration enthalpy increases as the radius of the ion increases.

TABLE 3-4: The molar enthalpy of hydration for the halide ions [7].

ION	$\Delta H_{Hydr.}$
F^-	−98 kcal/mol
Cl^-	−79 kcal/mol
Br^-	−71 kcal/mol
I^-	−62 kcal/mol

Large ions, such as iodide, have a high affinity for the stationary phase of an anion exchange resin. Since the hydration enthalpy of these ions is partly compensated by the cavity-building enthalpy, such ions will be described hereafter as hydro*phobic*. In contrast, smaller ions will be described as hydro-*philic*. This differentiation will be made in the following discussion, although all these ions are generally considered as hydrophilic when compared to some nonpolar organics.

Survey

With the separators, HPIC-AS1 to AS4A, it is possible to separate a wide range of inorganic anions such as:

Halides:
F^-, Cl^-, and Br^-

Oxygen containing halides:
OCl^-, ClO_2^-, ClO_3^-, BrO_3^-, and IO_3^-

Oxygen containing phosphorous anions:
HPO_2^{2-}, HPO_3^{2-}, and HPO_4^{2-}

Sulfur anions:
HS^-, SO_3^{2-}, and SO_4^{2-}

Nitrogen containing anions:
N_3^-, NO_2^-, NO_3^-, and CN^-

Silicon containing anions:
$H_2SiO_3^-$, and SiF_6^{2-}

Other non-metal oxy anions:
AsO_3^{3-}, AsO_4^{3-}, SeO_3^{2-}, and SeO_4^{2-}.

Besides the ions listed above, other nonmetallic anions, which are of research interest, can be separated by ion chromatographic techniques. The majority of the ions listed above can be separated using the standard $NaHCO_3/Na_2CO_3$ solution as eluant.

For the stationary phase of an anion exchange resin, the affinity of these ions is determined by two, solute-specific properties:

1. The Valency of the Sample Ion

The greater the valency of the ion, the longer is its retention time. For example, the monovalent nitrate ion elutes before the divalent sulphate. An exception to this rule is provided by the orthophosphate ion, which elutes under standard conditions as HPO_4^{2-} before the bromide. The elution of phosphate ions is very dependent on the eluant pH which controls the various dissociation equilibration. This point will be explained in more detail in the next chapter.

2. Ionic Radii in the Hydrated State

The retention time for an ion increases as the size of the ion in the hydrated state increases and with the resulting polarizability of the ion. Thus, halide ions elute in the following order: Fluoride, chloride, bromide and iodide. Indeed, the difference in retention times for bromide and iodide is so large, that the latter cannot be analyzed using a standard eluant (see Chapter 3, Section 2, "Hydrophobic Anions"). The size of an ion is generally more important for its retention time than its valency. For example, sulphate elutes before the monovalent—but highly polarizable—thiocyanate ion.

Experimental Retention Parameters

Choice of Eluant: The choice of eluant, or eluant composition, based on the list given in Table 3-2 depends, of course, on the nature of the ions to be separated. In general, the eluant should have a similar affinity for the stationary phase as the ions to be analyzed. For the majority of applications, a mixture of sodium carbonate and sodium bicarbonate has proved to be useful, because the resulting selectivity can be varied by varying the proportions of carbonate and bicarbonate. As can be seen from Figure 3-10, both mono- and divalent inorganic anions can be separated using this eluant mixture.

A better resolution in the region close to the void volume can be achieved by using more dilute ($0.7 - 1.5 \cdot 10^{-3}$ m) bicarbonate solutions as eluant. Such eluants are particularly useful for separating mixtures of, for example, BrO_3^-/Cl^- and $F^-/HCOO^-$. Since the dilute bicarbonate solutions are very weak eluants, it should be noted that subsequent analyses may be disturbed by the later eluting nitrate and sulphate ions, should these be present in the sample.

In addition, ions—such as sulfide and cyanide—also elute near the void volume; but these can only be detected amperometrically. Thus, the different detection systems help to improve the selectivity of the method. If the column effluent is allowed to flow through an amperometric cell before it flows through a fiber suppressor, a simultaneous detection of these ions is possible (see Figure 3-11).

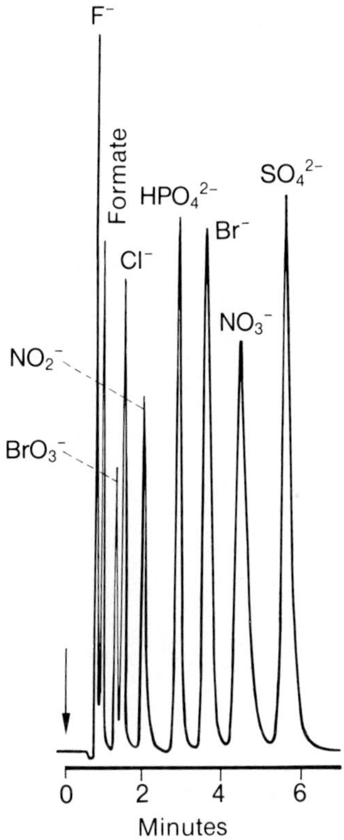

FIGURE 3-10.
A chromatogram of some inorganic mineral acids. **Eluant:** 0.0028 mol/dm^3 $NaHCO_3$ and 0.0022 mol/dm^3 Na_2CO_3. **Flow rate:** 2 cm^3/min. **Detection:** Suppressed conductivity. **Sample:** 50 μl anion standard.

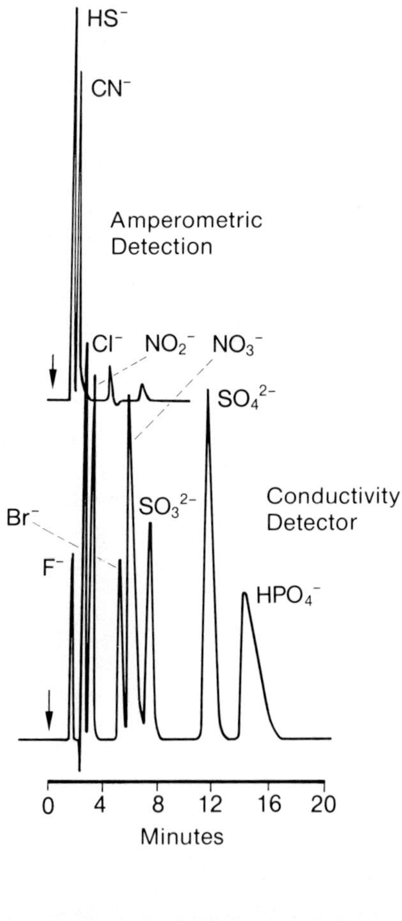

FIGURE 3-11.
Simultaneous detection of some inorganic ions by conductivity and amperometric sensors. **Column:** HPIC-AS3. **Eluant:** 0.001 mol/dm^3 Na_2CO_3, 0.01 mol/dm^3 NaH_2BO_3 and 0.015 mol/dm^3 ethylene diamine. **Flow rate:** 2.3 cm^3/min. **Detection:** Suppressed conductivity and amperometric with a silver working electrode.

Postcolumn derivatization, followed by photometric detection, can also be run simultaneously with suppressed conductivity detection. As shown in Figure 3-12, this kind of technique is essential for the analysis of water soluble silicate.

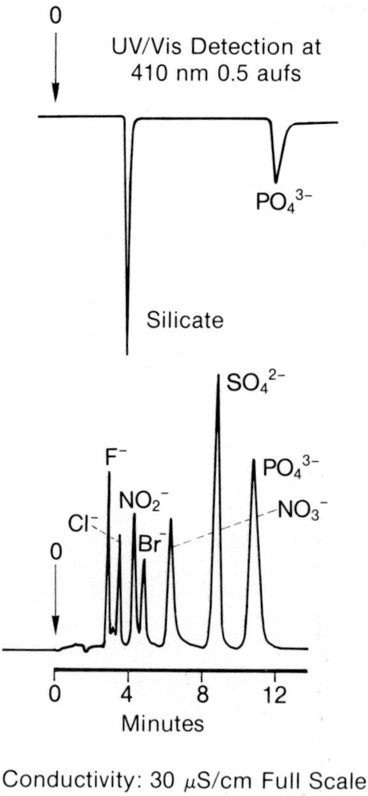

FIGURE 3-12.
Simultaneous conductivity and amperometric detection of some mineral acids and silicate. **Column:** MPIC-NS1. **Eluant:** 0.001 mol/dm^3 tetrapropylammonium hydroxide and 0.001 mol/dm^3 Na_2CO_3. **Detection:** Conductivity with suppression technique and photometrically after reaction with ammonium molybdate (λ = 410 nm).

A better resolution of the first part of the chromatogram shown in Figure 3-10 can be achieved by using a tetraborate eluant. This particularly weak eluant is only used for the separation of fluoride ions from aliphatic monocarboxylic acids having a small number of carbon atoms. (When preparing this eluant, it is important that the deionized water is thoroughly bubbled with an inert gas in order to ensure that the water is free of dissolved carbon dioxide.)

A variation in the selectivity in the elution range between phosphate and sulfate can be achieved by varying the concentration ratio of sodium carbonate/bicarbonate. In this region, ions such as sulfite, chlorate, and the oxygen-containing arsenic and selenic anions also elute.

The separator HPIC-AS2 was developed for the analysis of inorganic mineral acids in the presence of high concentrations of nitrate. As a result of the different functional groups on the latex particles, sulfate elutes before bromide and nitrate. As can be seen from Figure 3-13, these columns allow even small amounts of chloride, phosphate, and sulfate to be analyzed in the presence of high concentrations of nitrate.

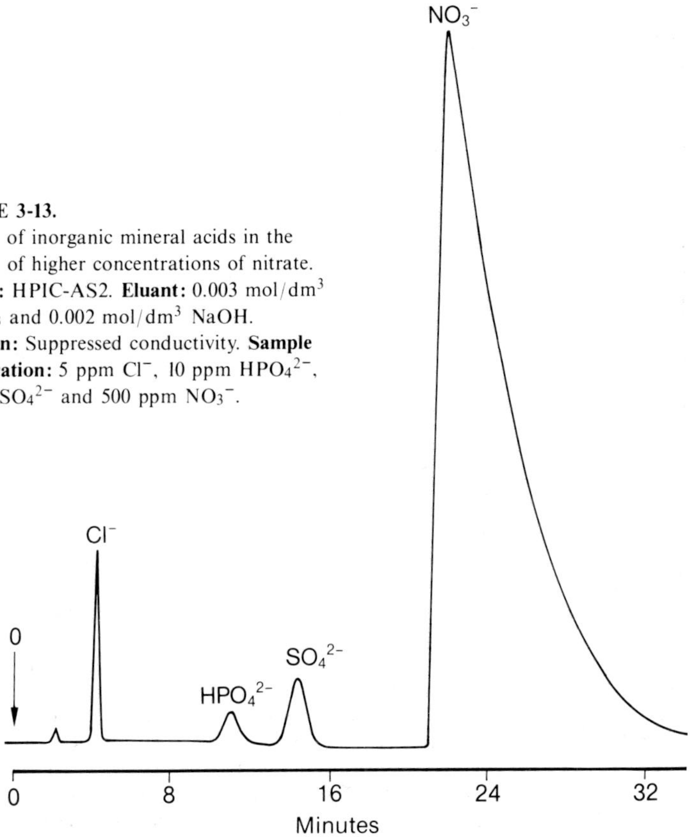

FIGURE 3-13.
Analysis of inorganic mineral acids in the presence of higher concentrations of nitrate. **Column:** HPIC-AS2. **Eluant:** 0.003 mol/dm^3 Na_2CO_3 and 0.002 mol/dm^3 NaOH. **Detection:** Suppressed conductivity. **Sample concentration:** 5 ppm Cl^-, 10 ppm HPO_4^{2-}, 10 ppm SO_4^{2-} and 500 ppm NO_3^-.

Optimal separations can be achieved using a sodium carbonate/sodium hydroxide eluant. A particularly important application with the AS2 separator is the separation of sulfate, bromide, and nitrate in concentrated brine solutions (see Figure 3-14). Separations of this type, where the component in excess elutes before the components to be analyzed, are only possible with AS2 or AS5 separators with their especially high exchange capacities. For these analyses, a mixture of sodium carbonate and tetrabutylammonium hydroxide is used as eluant; and, in order to improve the shape of the bromide and nitrate peaks, a little p-cyanophenol is added. This organic additive blocks the adsorption sites on the surface of the stationary phase and reduces the tailing factor.

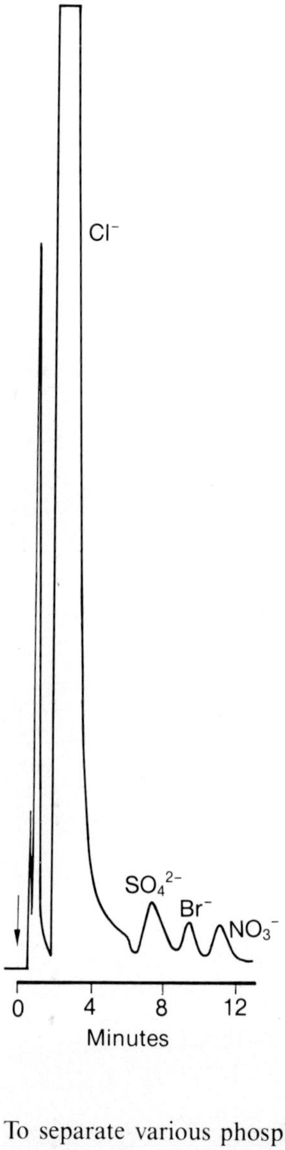

FIGURE 3-14.
Analysis of bromide, sulphate and nitrate in the presence of higher concentrations of chloride. **Column:** HPIC-AS2. **Eluant:** 0.0045 mol/dm^3 Na_2CO_3, 0.002 mol/dm^3 TBAOH and 0.00075 mol/dm^3 p-cyanophenol. **Detection:** Suppressed conductivity. **Sample concentration:** 1% NaCl and 5 ppm SO_4^{2-}, NO_3^- and Br^-.

To separate various phosphorous species—such as hypophosphite, phosphite, and phosphate (see Figure 3-15)—a sodium carbonate eluant is used which allows such ions to be separated in a single run and detected via their conductivity.

In the past, oxygen containing metal anions have also been separated using a more concentrated sodium carbonate eluant. However, utilizing these techniques results in problems associated with suppressing the background conductivity. Hence, for such separations, stationary phases of a special kind have been developed (see Chapter 3, Section 2, "Hydrophobic Anions").

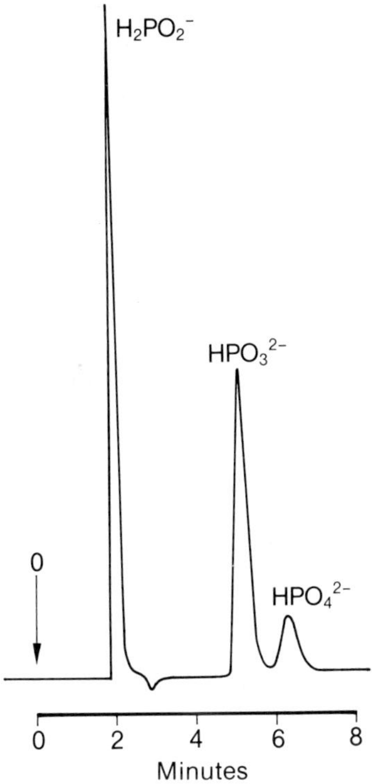

FIGURE 3-15.
The separation of phosphorous containing species such as hypophosphite, phosphite and phosphate. **Column:** HPIC-AS3. **Eluant:** 0.003 mol/dm^3 Na_2CO_3. **Flow rate:** 2.3 cm^3/min.

So far, the examples discussed emphasize that, by the appropriate selection of eluant and detection systems, it is possible to analyze not only ions which are present in similar concentrations, but also samples containing ions in very different concentrations. However, as with all chromatographic methods, ion chromatographic applications are limited by the difference between the retention times and concentrations of the ions being analyzed. Thus, chloride and sulphate anions, which are separated well, can be analyzed even in concentration ratios of 100,000:1; but for chloride and nitrite, the limiting ratio is only 500:1 using suppressed conductivity detection.

Eluant Concentration and pH: Eluant concentration and pH are two of the most important factors which determine the retention times of the ions being chromatographed. Thus, when considering changes in the selectivity of a system by varying the ratio of bicarbonate/carbonate, the ionic strength and the pH of the eluant must be treated simultaneously.

Whereas an increase in the eluant concentration decreases the retention of mono- and divalent ions, the pH of the eluant determines the retention of multivalent ions because their valency is a function of the eluant pH. This effect can be demonstrated by examining tribasic, orthophosphoric acid. The dissociation of this acid involves three steps [8]:

$$H_3PO_4 \rightleftharpoons H^+ + H_2PO_4^- \qquad pK_1 = 2.16 \qquad (26)$$

$$H_2PO_4^- \rightleftharpoons H^+ + HPO_4^{2-} \qquad pK_2 = 7.21 \qquad (27)$$

$$HPO_4^{2-} \rightleftharpoons H^+ + PO_4^{3-} \qquad pK_3 = 12.33 \qquad (28)$$

Due to these pK equilibria, an ion chromatographic separation of all three species is impossible. If sodium hydroxide is used as an eluant, the pH of the eluant rises with increasing concentration.

In Figure 3-16, a chromatogram obtained from injection of a sample containing F^-, Cl^-, NO_2^-, PO_4^{3-}, Br^-, NO_3^-, and SO_4^{2-} is shown. This chromatogram was obtained using a 0.001 mol/dm^3 sodium hydroxide eluant (pH 11). Note that phosphate and sulphate ions do not elute under these conditions. This is due to the fact that the concentration of the monovalent hydroxyl ions is not sufficient to elute the divalent sulphate and phosphate anions. If the sodium hydroxide concentration is increased, the retention times for the mono- and divalent ions, including sulfate, will decrease. However, at the resultant higher pH, the proportion of phosphate in the PO_4^{3-} form increases and, thus, the average retention time is also increased.

The system becomes even more complicated if a sodium bicarbonate/sodium carbonate mixture is used as eluant.

$$HCO_3^- \rightleftharpoons H^+ + CO_3^{2-} \qquad pK_2 = 10.3 \qquad (29)$$

Because of the equilibrium, the ratio of carbonate/bicarbonate will vary if the pH of the system is varied by addition of boric acid or sodium hydroxide. In order to examine this system in more detail, a standard mixture (0.0028 mol/dm^3 $NaHCO_3$; 0.0022 mol/dm^3 Na_2CO_3) was prepared and the bicarbonate concentration held constant while the carbonate concentration was varied between 0.5×10^{-3} mol/dm^3 and 2.22×10^{-3} mol/dm^3. In subsequent experiments, the carbonate concentration was held constant at 2.22×10^{-3} mol/dm^3 and the bicarbonate concentration varied. Under these conditions, the capacity factors of the seven inorganic anions listed above were measured. A plot of ln k′ versus the pH of the eluant, calculated from the various carbonate/bicarbonate concentrations according to

$$pH = pK_2 + \log \frac{[CO_3^{2-}]}{[HCO_3^-]} \qquad (30)$$

is shown in Figure 3-17.

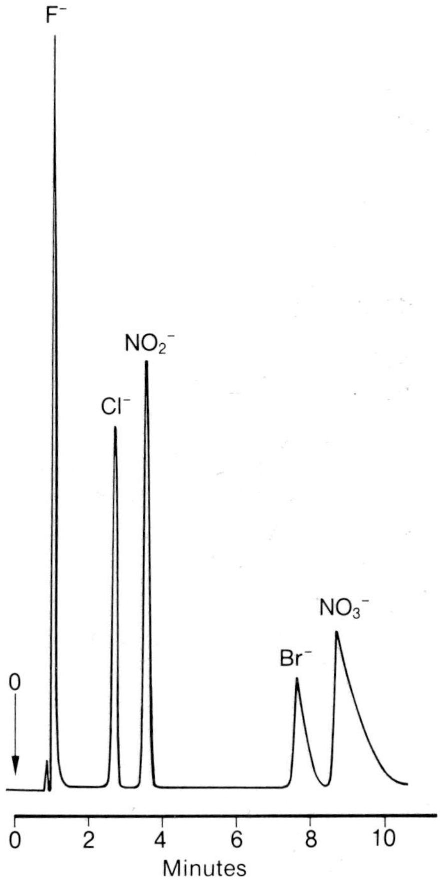

FIGURE 3-16.
Chromatogram from a separation of some mineral acids using NaOH as eluant. **Column:** HPIC-AS3. **Eluant:** 0.001 mol/dm^3 NaOH. **Flow rate:** 2.3 cm^3/min. **Detection:** Suppressed conductivity. **Sample:** 50 μl anion standard.

Figure 3-17 shows that the retention times of all ions decrease as the concentration of carbonate increases (the concentration of bicarbonate was held constant; pH range 8.35 – 10.23). On the other hand, if the concentration of bicarbonate is increased and that of the carbonate held constant (pH range: 10.98 – 10.28), the retention times do not decrease significantly. The sulphate ions are most affected by these variations and this can be explained by the ineffectiveness of the monovalent bicarbonate as an eluent anion. Also visible in Figure 3-17 is that the changing pH of the eluant mixtures, brought about by the changing concentration ratio of the carbonate species, leads to a large variation in the retention time for the phosphate anion; so that, in a very narrow pH range (pH 9.6 – 10.0), the bromide and nitrate peaks become superimposed on that of the phosphate anion. Even at pH > 10.8, these peaks interfere with one another. These effects can be explained in terms of the position of the phosphate dissociation equilibrium discussed earlier.

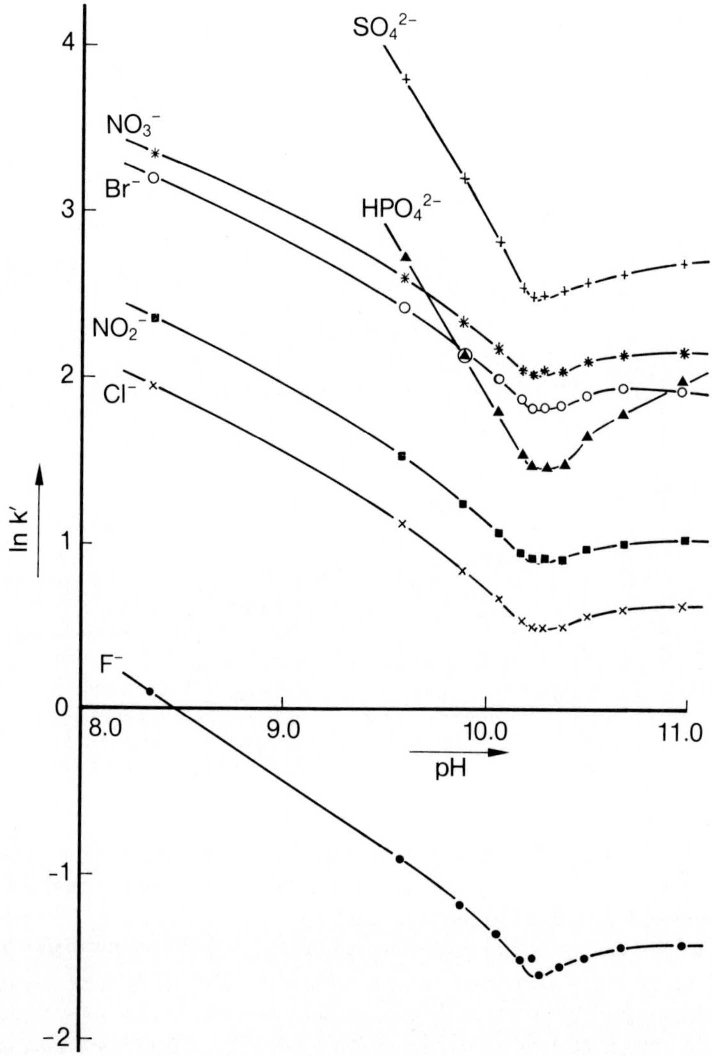

FIGURE 3-17.
The dependence of retention time on the pH of the eluant (adjusted by varying the carbonate/bicarbonate ratio) for some inorganic mineral acids. **Column:** HPIC-AS4. **Flow rate:** 2 cm^3/min. **Detection:** Suppressed conductivity. **Sample:** 50 μl anion standard.

DATA FOR FIGURE 3-17.

C_{Eluant}	pH	ln k′						
		F^-	Cl^-	NO_2^-	HPO_4^{2-}	Br^-	NO_3^-	SO_4^{2-}
2.8 mM HCO_3^-	8.35	0.11	1.96	2.35	—	3.19	3.35	—
2.8 mM HCO_3^- 0.5 mM CO_3^{2-}	9.58	−0.89	1.13	1.52	2.72	2.42	2.60	3.79
2.8 mM HCO_3^- 1.0 mM CO_3^{2-}	9.88	−1.17	0.84	1.23	2.13	2.11	2.33	3.19
2.8 mM HCO_3^- 1.5 mM CO_3^{2-}	10.06	−1.35	0.67	1.07	1.79	1.99	2.17	2.82
2.8 mM HCO_3^- 2.0 mM CO_3^{2-}	10.18	−1.51	0.54	0.94	1.53	1.87	2.05	2.55
2.8 mM HCO_3^- 2.22 mM CO_3^{2-}	10.23	−1.51	0.51	0.91	1.47	1.83	2.02	2.48
C_{Eluant}	pH	ln k′						
		F^-	Cl^-	NO_2^-	HPO_4^{2-}	Br^-	NO_3^-	SO_4^{2-}
2.22 mM CO_3^{2-} 0.5 mM HCO_3^-	10.98	−1.43	0.63	1.03	1.94	1.93	2.16	2.69
2.22 mM CO_3^{2-} 1.0 mM HCO_3^-	10.68	−1.43	0.61	1.00	1.78	1.94	2.14	2.62
2.22 mM CO_3^{2-} 1.5 mM HCO_3^-	10.50	−1.51	0.57	0.97	1.64	1.90	2.10	2.58
2.22 mM CO_3^{2-} 2.0 mM HCO_3^-	10.38	−1.56	0.51	0.90	1.48	1.84	2.03	2.44
2.22 mM CO_3^{2-} 2.5 mM HCO_3^-	10.28	−1.61	0.51	0.92	1.48	1.84	2.04	2.48

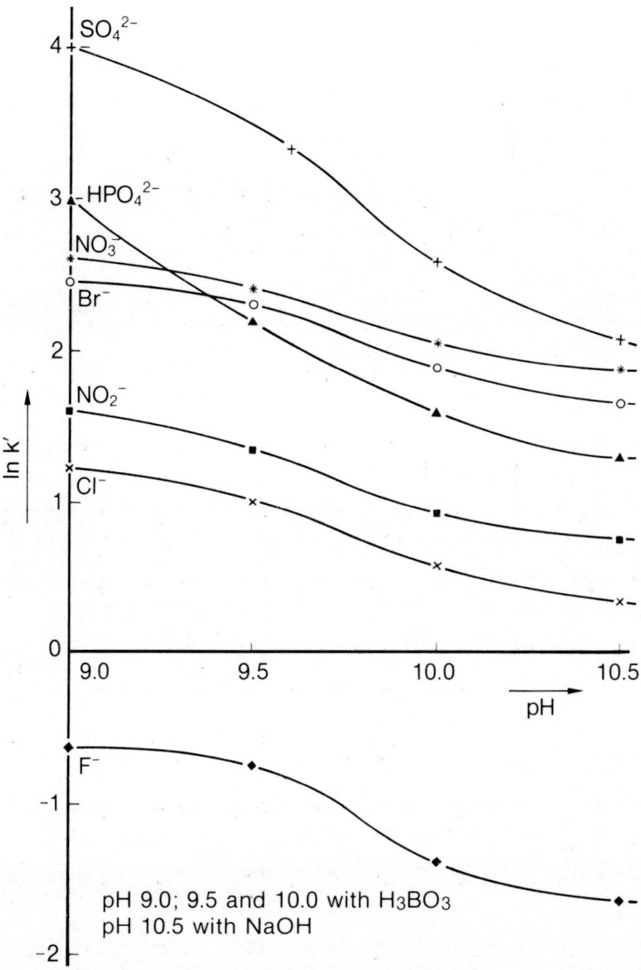

FIGURE 3-18.
The dependence of retention time on the pH of the buffer solution at constant ionic strength. **Column:** HPIC-AS3. **Eluant:** 0.0028 mol/dm^3 $NaHCO_3$ and 0.0022 Na_2CO_3; pH adjusted with H_3BO_4 and NaOH. **Detection:** Suppressed conductivity. **Sample:** 50 μl anion standard.

DATA FOR FIGURE 3-18.

	t_{ms} [min]	k′	ln k′	t_{ms} [min]	k′	ln k′	t_{ms} [min]	k′	ln k′	t_{ms} [min]	k′	ln k′
F^-	1.30	0.53	−0.63	1.26	0.48	−0.73	1.06	0.25	−1.39	1.01	0.19	−1.66
Cl^-	3.73	3.39	1.22	3.16	2.72	1.00	2.35	1.76	0.57	2.05	1.41	0.34
NO_2^-	5.00	4.88	1.59	4.17	3.91	1.36	3.06	2.50	0.92	2.64	2.11	0.75
HPO_4^{2-}	17.55	19.65	2.98	8.48	8.98	2.19	5.00	4.88	1.59	3.97	3.67	1.30
Br^-	10.61	11.48	2.44	9.36	10.01	2.30	6.38	6.51	1.87	5.40	5.35	1.68
NO_3^-	12.38	13.56	2.61	10.22	11.02	2.40	7.53	7.86	2.06	6.35	6.47	1.87
SO_4^{2-}	47.35	54.71	4.00	24.52	27.85	3.33	12.18	13.33	2.59	7.60	7.94	2.07
pH	←— 9.0 —→			←— 9.5 —→			←— 10.0 —→			←— 10.5 —→		

Separator: HPIC-AS3 Flow Rate: 2.3 cm^3/min

$c_{NaHCO_3} = 0.0028$ mol/dm^3 $c_{Na_2CO_3} = 0.022$ mol/dm^3

pH adjusted with $H_3BO_3/NaOH$

If the ionic strength of the eluant is held constant and the pH is varied by adding, for example, boric acid or sodium hydroxide, then the retention times of inorganic anions vary as shown in Figure 3-18. As expected, the retention times decrease as the pH increases. Thus, by addition of boric acid, the dissociation equilibrium of the carbonate is shifted towards the hydrogen carbonate side and, since hydrogen carbonate is the weaker eluant, the retention times increase. In this case, interference between the phosphate peak and those of the bromide and nitrate anions occurs between pH 9.0 and 9.5.

From the examples discussed above and shown in Figures 3-17 and 3-18, it is clear that the pH of the eluant cannot be varied without changing the carbonate/bicarbonate equilibrium and, thus, the effect of pH cannot be investigated independently. However, if the carbonate/bicarbonate concentration ratio is held constant and the ionic strength of the eluant is varied, it is possible, from the variation in the retention times, to deduce the effects of changes in the pH of the eluant. Thus, a series of experiments were conducted where the total carbonate concentration was varied between 3 and 6×10^{-3} mol/dm^3 while keeping the concentration ratio of carbonate/bicarbonate constant. A plot of the ln k′ values so obtained as a function of the ionic strength is shown in Figure 3-19. At a pH of 10.28, which results from the particular carbonate/bicarbonate ratio used, the sulphate ion is not the only divalent anion to elute. The phosphate anions also elute in their HPO_4^{2-} form. As expected, the retention times for the divalent ions are more influenced by changes in the ionic strength than are the monovalent ions. If the dependence of the retention times for the bromide, nitrate, and phosphate ions is extrapolated to lower ionic strengths, it can be seen that the peaks resulting from these ions will overlap when the ionic

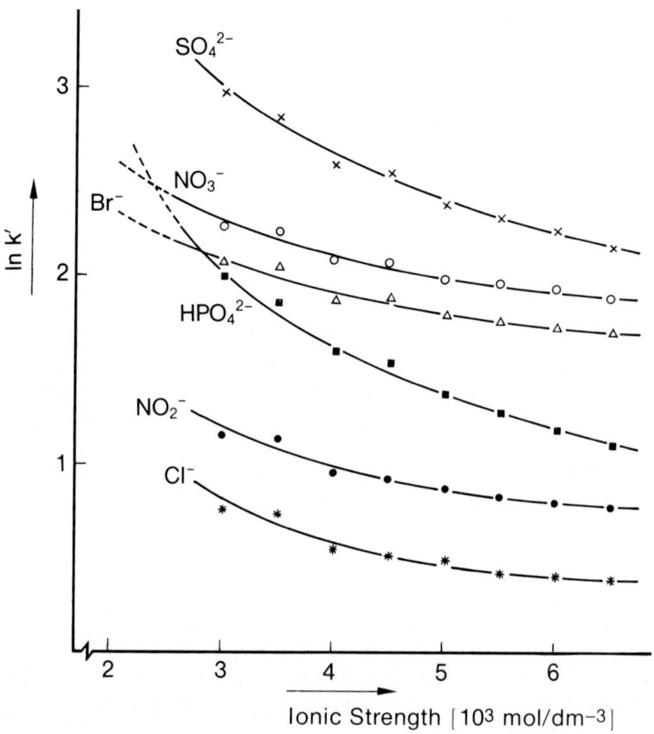

FIGURE 3-19.
The influence of ionic strength on the retention times for some inorganic mineral acids at constant pH. **Column:** HPIC-AS4. **Flow rate:** 2 cm^3/min. **Detection:** Suppressed conductivity. **Sample:** 50 μl anion standard.

DATA FOR FIGURE 3-19.

Concentration	ln k′						
10^3 mol/dm^3	F^-	Cl^-	NO_2^-	HPO_4^{2-}	Br^-	NO_3^-	SO_4^{2-}
3.02	−1.26	0.76	1.15	2.00	2.07	2.27	2.97
3.52	−1.11	0.74	1.13	1.86	2.05	2.23	2.84
4.02	−1.50	0.55	0.96	1.60	1.89	2.08	2.59
4.52	−1.55	0.51	0.92	1.55	1.88	2.07	2.55
5.02	−1.35	0.49	0.88	1.37	1.80	1.98	2.38
5.52	−1.61	0.42	0.83	1.28	1.77	1.96	2.31
6.02	−1.50	0.40	0.80	1.18	1.73	1.93	2.23
6.52	−1.61	0.38	0.77	1.10	1.70	1.88	2.14

strength is between 2 and 3 mol/dm^3. Furthermore, at much higher ionic strengths than those investigated, a change in the elution sequence for sulphate and nitrate can be predicted.

A comparison of Figures 3-18 and 3-19 suggests that the retention times for inorganic anions are essentially determined by the ionic strength of the eluant. Changes in the pH of the medium at constant ionic strength do lead to a shifting of the carbonate/bicarbonate equilibrium, but this affects the retention times only indirectly.

Eluant Flow Rate: Just as for silica-based HPLC columns, the van-Deemter curve for an anion exchange column of the type HPIC-AS4 (shown in Figure 2-6), shows how the theoretical plate height varies considerably only at low flow rates as the flow rate is altered. At higher flow rates, the HETP is almost independent of the flow rate. Thus, by increasing the flow rate, the analysis times can be shortened without significant loss of separation capability. The flow rate and the retention times are directly proportional to one another. The flow rate cannot, however, be increased indefinitely. The limit is set by the backpressure of the column and by the pressure compatibility of the membrane suppressor. Furthermore, the ability to increase the resolution by reducing the flow rate is limited. Thus, since the flow rate does not affect the pH and ionic strength of the eluant, but does control the elution speed of the ions, the retention times are most easily regulated through this parameter.

Column Length: Other parameters being equal, the length of the column determines the number of theoretical plates and, thus, the degree of separation which the column can achieve. If two columns are used in series, the exchange capacity of the system will increase and, as a consequence, the retention times become longer. An improvement in the separation capability of the system leads to better resolution of ions having similar retention times. Using longer columns; i.e., increasing the exchange capacity of the columns, is also recommended when an analysis must be made in the presence of an excess of one particular ion.

Using shorter columns is rarely worthwhile since excess resolution; i.e., excessively long analysis times, between two analytes can more easily be avoided by an appropriate choice of eluant and column type. However in some exceptional cases, retention times can be drastically reduced by using a precolumn instead of a full column.

Thermodynamic Aspects: The way in which an ion will be retained on a column can be formulated in terms of the capacity factor k′:

$$k'_i = \frac{t^i_{ms} - t^i_m}{t^i_m} \tag{31}$$

The capacity factor is directly related to the distribution coefficient K and, through this, with the thermodynamic quantities $\Delta H_{m\to s}$ (enthalpy) and $\Delta S_{m\to s}$ (entropy) which determine the chromatographic process. (The subscript m→s denotes the process of material transfer from the **m**obile to the **s**tationary phases.)

$$k'_i = K_i \cdot \Phi = K_i \cdot \frac{V_{stat}}{V_{mob}} \qquad (32)$$

Φ: The ratio of the volumes of the two phases.
V_{stat}: Volume of the stationary phase.
V_{mob}: Volume of the mobile phase; i.e., the dead volume.

The retention of a solute A as a function of its $\Delta G_{m\to s}$ (A) and the ratio of the volumes of the two phases can be described thermodynamically [9, 10 and 11] by:

$$\ln k' = -\frac{\Delta G_{m\to s}\,(A)}{RT} + \ln \Phi \qquad (33)$$

If A is a member of an homologous, or of a quasi-homologous series, then:

$$\Delta G_{m\to s} = \Delta G^*_{m\to s} + \Delta\Delta G_{m\to s} \cdot n \qquad (34)$$

where $\Delta\Delta G_{m\to s}$ is the change in the free energy of sorption with each progression in the series. The term $\Delta G^*_{m\to s}$ allows for the nonlinearity of $\Delta G_{m\to s}$ with n for the smallest member of each of the series and also incorporates a contribution for the functional group. From this treatment, it follows that:

$$\ln k' = -\frac{\Delta G^*_{m\to s}}{RT} + \ln \Phi - \frac{\Delta\Delta G_{m\to s}}{RT} \cdot n \qquad (35)$$

By plotting ln k′ as a function of n, a straight line is obtained for each homologous series, so that:

$$\ln k' = a + b \cdot n \qquad (36)$$

From Equation 35, it follows that a and b are given by:

$$a = -\frac{\Delta G^*_{m\to s}}{RT} + \ln \Phi \qquad (37)$$

$$b = -\frac{\Delta\Delta G_{m\to s}}{RT} \qquad (38)$$

Since

$$\Delta G = \Delta H - T \cdot \Delta S \qquad (39)$$

Equations 37 and 38 can be rewritten in the forms:

$$a = -\frac{\Delta H^*_{m\rightarrow s}}{RT} + \frac{\Delta S^*_{m\rightarrow s}}{R} + \ln \Phi \qquad (40)$$

$$b = -\frac{\Delta\Delta H_{m\rightarrow s}}{RT} + \frac{\Delta\Delta S_{m\rightarrow s}}{R} \qquad (41)$$

and Equation 35 can now be rewritten in the form:

$$\ln k' = -\frac{\Delta H^*_{m\rightarrow s}}{RT} + \frac{\Delta S_{m\rightarrow s}}{R} - \left(\frac{\Delta\Delta H_{m\rightarrow s}}{RT} - \frac{\Delta\Delta S_{m\rightarrow s}}{R}\right) \cdot N + \ln \Phi \qquad (42)$$

Values of $\ln k'$ and $\Delta H_{m\rightarrow s}$ can be obtained experimentally from the temperature dependence of the retention times. To derive values for the adsorption entropy terms, the volume ratio of the two phases must be known. Since the capacity factor can be directly correlated with $\Delta G_{m\rightarrow s}$ via the distribution coefficient K (Equation 32), and, also because the sorption processes involve both distribution and adsorption, the thermodynamic data can only be treated as formal quantities.

The basic model developed by Eon and Guiochon [12, 13] for treating the adsorption effects at gas/liquid and liquid/solid phase boundaries has been modified by Möckel [11] for describing the problem of retention on chemically bound reversed phases in HPLC. This model has only limited applicability to ion chromatographic processes.

Nevertheless, initial attempts to thermodynamically characterize the retention behavior of inorganic anions [14] do suggest some similarities to HPLC. In this work, the retention times for seven anions were studied as a function of column temperature T. A plot of $\ln k'$ against $1/T$ (van't Hoff plot) for the nitrate and bromide ions is shown in Figure 3-20. In contrast, over the temperature range studied, the retention of the fluoride, chloride, nitrite, phosphate, and sulphate ions was—within experimental error limits—independent of temperature. This behavior is understandable in as much as the bromide and nitrate ions can be separated simply on the basis of their different adsorption properties. Treating the data given in Table 3-5 with a linear regression leads to the following equation:

$$\text{Bromide: } \ln k' = 1{,}036.03 \cdot T^{-1} - 2.56 \ ; \ r = 0.994 \qquad (43)$$

$$\text{Nitrate: } \ln k' = 968.38 \cdot T^{-1} - 2.17 \ ; \ r = 0.983 \qquad (44)$$

From the slope of the straight lines, the sorption enthalpy $\Delta H_{m\rightarrow s}$ can be obtained using:

$$b = -\frac{\Delta H_{m\rightarrow s}}{R} \qquad (45)$$

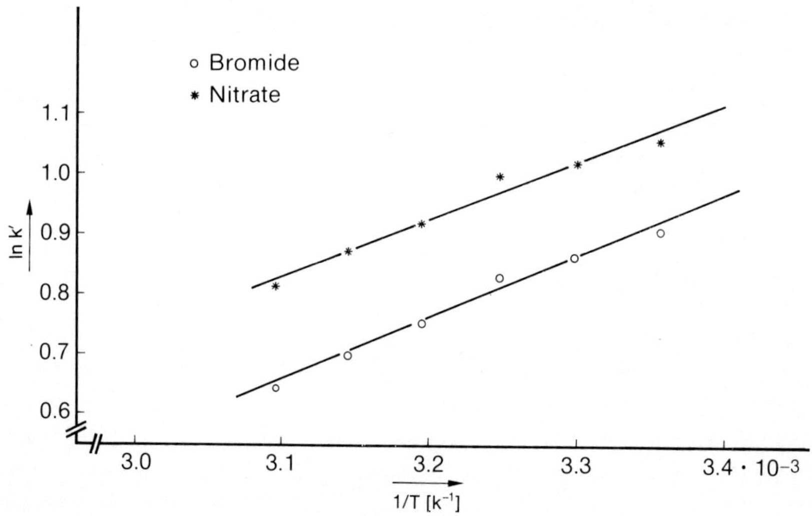

FIGURE 3-20.
Variation of the retention of bromide and nitrate ions with temperature. **Column:** HPIC-AS3. **Eluant:** 0.0028 mol/dm^3 $NaHCO_3$ and 0.0022 mol/dm^3 Na_2CO_3. **Flow rate:** 2.3 cm^3/min. **Detection:** Suppressed conductivity. **Sample:** 50 μl anion standard.

TABLE 3-5: ln k′ values at different temperatures for the bromide and nitrate anions.

T [K]	T^{-1} [K^{-1}]	ln k′ BROMIDE	ln k′ NITRATE
298	$3.356 \cdot 10^{-3}$	0.909	1.060
303	$3.300 \cdot 10^{-3}$	0.862	1.019
308	$3.247 \cdot 10^{-3}$	0.829	1.004
313	$3.195 \cdot 10^{-3}$	0.752	0.918
318	$3.145 \cdot 10^{-3}$	0.699	0.871
323	$3.096 \cdot 10^{-3}$	0.645	0.813

With R = 8.313 J/(K·mol), the sorption enthalpies for these two ions are found to be:

Bromide: $\Delta H_{m \to s} = -8.6 \text{ kJ mol}^{-1} \cdot K^{-1}$

Nitrate: $\Delta H_{m \to s} = -8.1 \text{ kJ mol}^{-1} \cdot K^{-1}$

These values are of the same order of magnitude as those found by Möckel [11] for nonionic organic and inorganic solutes. The sorption entropy, which is usually obtained from the intercept of the ln k′ plots using

$$a = \frac{\Delta S_{m \to s}}{R} + \ln \Phi \qquad (46)$$

could not be calculated for the example discussed above because the phase volume ratio was not known. The latest work from Möckel [11] suggests that the phase volume ratio can be easily determined for reversed phase chromatography. Thus, the capacity factors for the individual members of an homologous series are measured using two eluant compositions and the values plotted as a function of n. Since for solutes for which K = 1 the free sorption energy $\Delta G_{m \to s}$ = 0, the phase volume ratio can then be found using

$$\ln k' = \ln \Phi \tag{47}$$

from the intercept of the two straight lines.

Before a similar treatment can be applied to an ion chromatography column, a suitable system will have to be found.

Since the values of ln k′ for the anions fluoride, chloride, nitrite, phosphate, and sulphate show no dependence on 1/T, the sorption enthalpy $\Delta H_{m \to s} = 0$ for these ions. Thus, any change in free sorption energy with temperature is derived solely from a change in the entropy of the process. Further experiments could show to what extent changes in the entropy of mixing—in the degree of order of water due to formation and dissolution of hydration spheres and the coordination entropy of the matrix—contribute to the molar sorption entropy.

Hydrophobic Anions

The group of hydrophobic inorganic anions includes iodide, thiocyanate, thiosulphate, and the metal oxide anions—chromate, molybdate, tungstate etc. Since (due to their large radii) these ions have considerable affinity for the stationary phase, they were (before the introduction of the AS5 columns) separated, using columns AS1 to AS4 and more concentrated ($6 - 8 \cdot 10^{-3}$ mol/dm^3 Na_2CO_3) carbonate eluants. The high carbonate concentration meant that the modern membrane suppressors could not be used to reduce the background conductivity, and the more traditional suppressor systems had the disadvantage that they required frequent regeneration. Strong tailing effects, which could be attributed to adsorption effects, limited the sensitivity of the analyses to concentrations above the ppm range.

The development of the AS5 columns has made the analysis of hydrophobic inorganic anions considerably easier. The functional groups on the latex particles in the AS5 columns are less hydrophobic and, thus, the separation of hydrophobic anions can be carried out with standard carbonate/hydrogen carbonate mixtures. In order to minimize adsorption effects, a small quantity of p-cyanophenol can be added to the eluant.

The effect of p-cyanophenol on the shape of the peaks obtained can be clearly seen from Figure 3-5. The compatibility of these newer systems with membrane suppressors and the associated reduction in dead volume, have greatly improved the degree of peak broadening. Further examples of chromatographic separations of hydrophobic anions are shown in Figures 3-21 and 3-22.

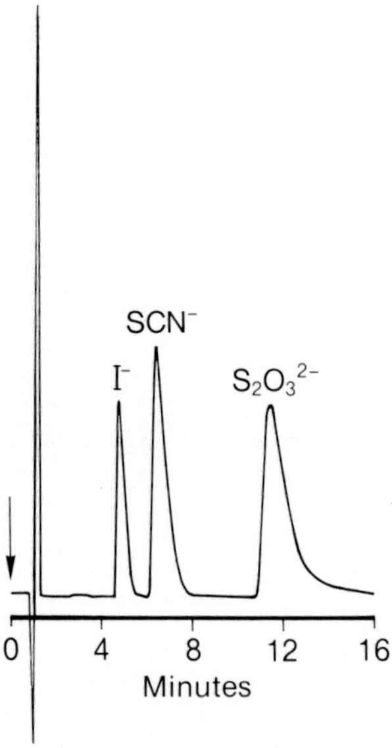

FIGURE 3-21.
The separation of iodide, thiocyanate and thiosulphate using an HPIC-AS5 column. **Eluant:** 0.0043 mol/dm^3 $NaHCO_3$, 0.0034 mol/dm^3 Na_2CO_3 and 100 mg/dm^3 p-cyanophenol. **Flow rate:** 2 cm^3/min. **Detection:** Suppressed conductivity. **Sample concentration:** 100 ppm I^-, SCN^- and $S_2O_3^{2-}$.

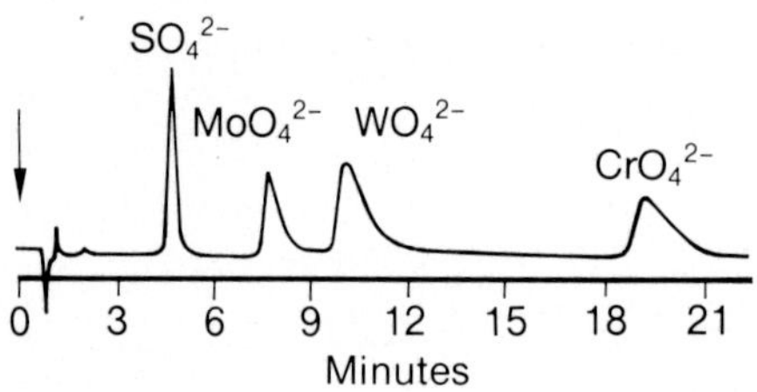

FIGURE 3-22.
The separation of oxygen containing metal anions. **Column:** HPIC-AS5. **Eluant:** 0.0043 mol/dm^3 $NaHCO_3$, 0.0034 mol/dm^3 Na_2CO_3 and 100 mg/dm^3 p-cyanophenol. **Flow rate:** 2 cm^3/min. **Detection:** Suppressed conductivity.

In analogy to the hydrophilic anions, the controlling factor for the retention times of the hydrophobic ions is the concentration of carbonate/hydrogen carbonate in the eluant. As the concentration is increased, the retention time of the chromate anion, for example, is considerably reduced (see Figure 3-22).

A further important application of these columns is analyses in the presence of excess chloride. An extreme example of such an analysis is shown in Figure 3-23.

The impressive separation capability of these columns is demonstrated clearly by the examples shown in Figures 3-24 and 3-25. Such analyses of complexes of transition and noble metals open a completely new area of application for ion chromatography.

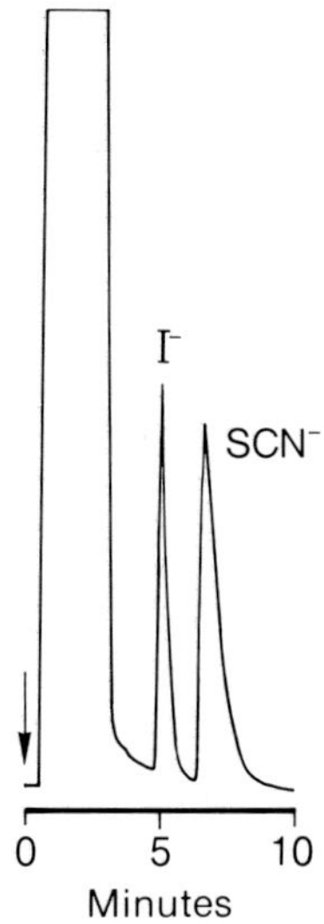

FIGURE 3-23.
Analysis of iodide and thiocyanate in the presence of higher concentrations of chloride. **Column:** HPIC-AS5. **Eluant:** 0.0043 mol/dm^3 $NaHCO_3$, 0.0034 mol/dm^3 Na_2CO_3 and 100 mg/dm^3 p-cyanophenol. **Flow rate:** 2 cm^3/min. **Detection:** Suppressed conductivity. **Sample concentration:** 7% NaCl, 100 ppm I^- and SCN^-.

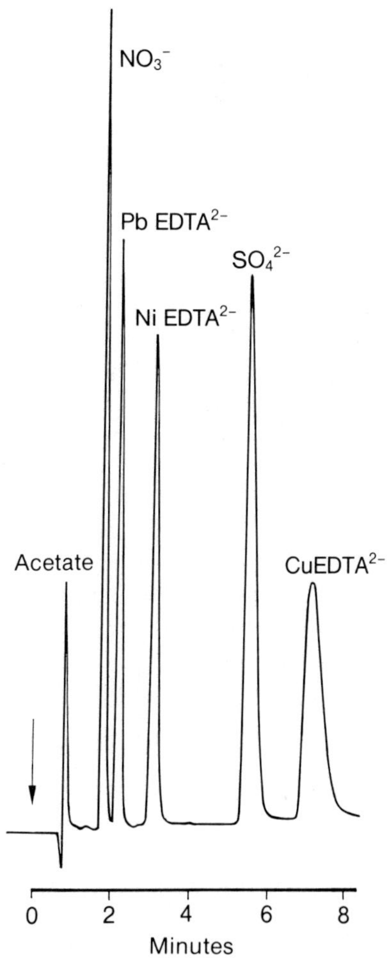

FIGURE 3-24.
The separation of some metal-EDTA complexes. **Column:** HPIC-AS5. **Eluant:** 0.002 mol/dm^3 Na_2CO_3 and 0.002 mol/dm^3 $NaHCO_3$. **Flow rate:** 2 cm^3/min. **Detection:** Suppressed conductivity.

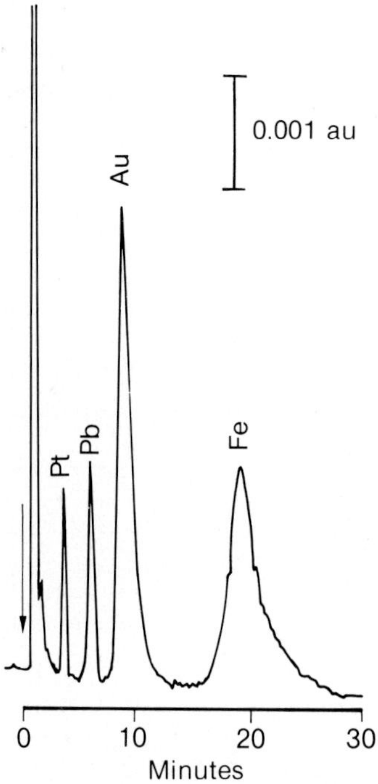

FIGURE 3-25.
The separation of some metal-chloro complexes. **Column:** HPIC-AS5. **Eluant:** 0.2 mol/dm^3 $NaClO_4$ and 0.2 mol/dm^3 HCl. **Flow rate:** 2 cm^3/min. **Detection:** UV at 254 nm.

Intensive research in this direction should soon produce a solution to the problem of separating chrome III and chrome VI anions by an analogous procedure.

Similar strong affinities for the stationary phase are exhibited by polyvalent inorganic anions. Most important among the polyvalent anions are the condensed phosphates — such as pyrophosphate and tripolyphosphate — which are of considerable interest to the soap industry. Ions such as these can be separated using an HPIC-AS7 column which was developed for such analyses. Figure 3-26 shows a typical chromatogram from a separation of pyrophosphate and tripolyphosphate.

The relatively concentrated nitric acid eluent reduces the degree of dissociation of the polyvalent phosphates and, thus, lowers their affinity for the stationary phase. In Figure 3-26, it can also be seen that — using postcolumn derivatization with iron (III) nitrate and a UV detector (λ = 330 nm) — other inorganic ions can also be detected. Hence, the orthophosphate and sulphate anions also react with iron to form complexes that can be detected during the same analysis. However, it must be emphasized that the determination of orthophosphate and sulphate, using conventional columns and a suppressed conductivity detection system, is much more sensitive.

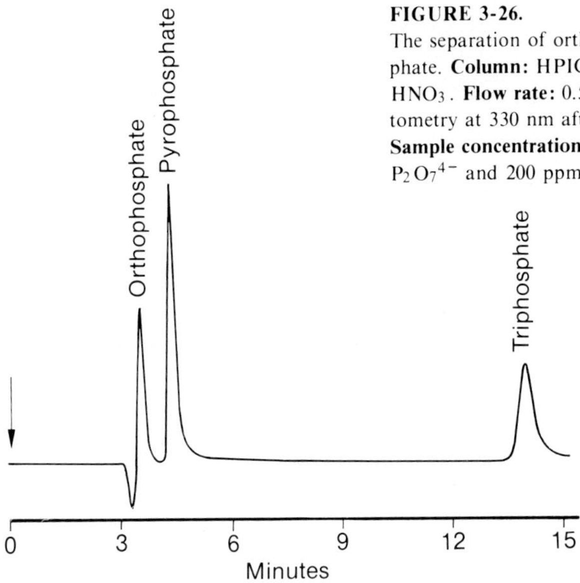

FIGURE 3-26.
The separation of ortho-, pyro- and tripolyphosphate. **Column:** HPIC-AS7. **Eluant:** 0.07 mol/dm^3 HNO_3. **Flow rate:** 0.5 cm^3/min. **Detection:** Photometry at 330 nm after reaction with $Fe(NO_3)_3$. **Sample concentration:** 100 ppm PO_4^{3-}, 50 ppm $P_2O_7^{4-}$ and 200 ppm $P_3O_{10}^{5-}$.

Several organic compounds can be analyzed using similar experimental conditions and an appropriate eluant concentration (see "Ion Exchange Chromatography of Organic Anions," later in this chapter). The concentration of the nitric acid is the only parameter affecting the retention times. An example of a chromatogram obtained from a separation of inorganic and organic compounds is shown in Figure 3-27.

ION EXCHANGE CHROMATOGRAPHY OF ORGANIC ANIONS

Organic Acids

It is also possible to analyze many organic species using the columns HPIC-AS1 to AS5 which, to some extent, exhibit retention behavior similar to that of the mineral acids. Thus, interferences are possible in samples containing both organic and inorganic anions.

Small aliphatic organic acids; e.g., acetate, propionate, elute close to the dead volume of the column and tend to make simultaneous analysis of fluoride difficult. If the "standard" eluant for the separation of mineral acids (0.0028 mol/dm^3 $NaHCO_3$ and 0.0022 mol/dm^3 Na_2CO_3) is used with the columns AS1 to AS4, it is possible to separate fluoride from formate. However, if the matrix also contains acetate, the separation (of fluoride and acetate) fails. This can often lead to an erroneous interpretation of the signal assigned to fluoride. Special

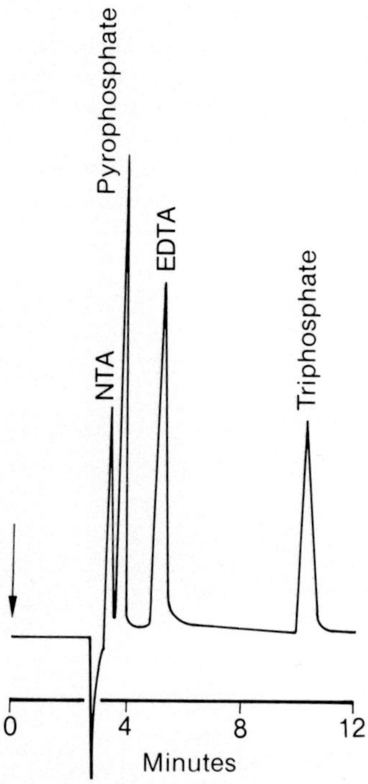

FIGURE 3-27.
The separation of inorganic anions and complexes. **Column:** HPIC-AS7. **Eluant:** 0.05 mol/dm^3 HNO_3. **Flow rate:** 0.5 cm^3/min. **Detection:** Photometry at 330 nm after reaction with $Fe(NO_3)_3$.

attention must be paid to environmental samples (such as wastewater, surface water, and digestion samples) in which fluoride can often be proved to be present but which also contain a large number of organic acids. Since the latter often coelute with fluoride, the interpretation of the first peak to elute can be extremely difficult. To determine the nature of the compounds responsible for this peak, a dilute bicarbonate eluant is recommended which, because of the low elution strength of bicarbonate, allows a separation of formate and acetate (see Figure 3-28). However, for the routine analysis of such complex matrices, this technique is not particularly useful because, even for ions such as chloride, the retention times become excessively long.

Ions, such as nitrate and sulphate, which elute even later, then interfere with subsequent analyses. For complex plant digests, the elution strength of bicarbonate is too great to achieve a separation of the organic acids and fluoride. In this case, the alternative eluant sodium tetraborate, which must be prepared with carbonate-free water (!), can be used. The separation of fluoride, acetate, and formate under such conditions led to the chromatogram shown in Figure 3-29. With these chromatographic conditions, chloride anions have retention times of ca. 12 minutes.

Benzoic acid elutes later than the aliphatic acids and has a retention time identical to that of nitrite. The only characteristic which can be used to distinguish between

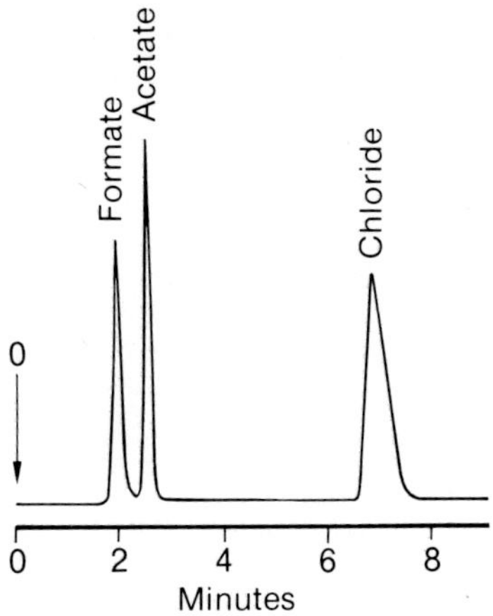

FIGURE 3-28.
The separation of formate, acetate and chloride. **Column:** HPIC-AS4. **Eluant:** 0.001 mol/dm^3 $NaHCO_3$. **Flow rate:** 2 cm^3/min. **Detection:** Suppressed conductivity. **Sample:** $2 \cdot 10^{-4}$ mol/dm^3 sodium formate, $5 \cdot 10^{-4}$ mol/dm^3 sodium acetate and 10 ppm Cl^-.

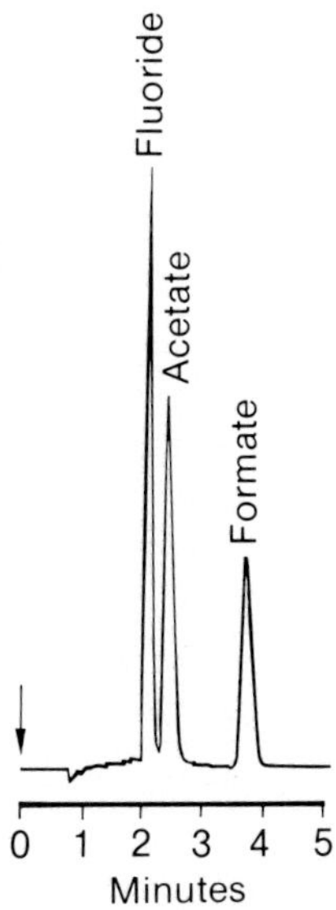

FIGURE 3-29.
The separation of formate, acetate and fluoride. **Column:** HPIC-AS4. **Eluant:** 0.002 mol/dm^3 $Na_2B_4O_7$. **Flow rate:** 2 cm^3/min. **Detection:** Suppressed conductivity. **Sample:** 3 ppm F^-, $5 \cdot 10^{-4}$ mol/dm^3 sodium acetate and $2 \cdot 10^{-4}$ mol/dm^3 sodium formate.

these two components is the peak shape. Whereas a nitrite peak is completely symmetrical, that of benzoic acid exhibits a small tailing effect due to adsorption of the aromatic ring on the stationary phase. Separation of nitrite and benzoic acid by ion chromatography has not yet been achieved. It is possible that this may be achieved using ion pair chromatography, which is the only available ion chromatographic method for separating nitrate and chlorate.

Since benzoic acid has a relatively short retention time, the separation of other aromatic monocarboxylic acids by anion exchange chromatography has been investigated. Figure 3-30 shows the chromatogram of a separation of mandelic acid, hippuric acid, and methyl substituted hippuric acids which, under standard conditions, is not very satisfactory.

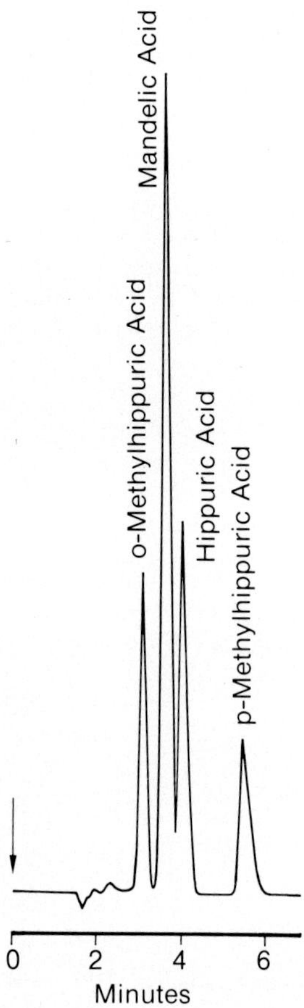

FIGURE 3-30.
The separation of aromatic monocarboxylic acids. **Column:** 2 HPIC-AS4. **Eluant:** 0.0005 mol/dm^3 Na_2CO_3 and 0.0005 mol/dm^3 NaOH. **Flow rate:** 2 cm^3/min. **Detection:** Suppressed conductivity. **Sample volume:** 50 μl.

However, using two HPIC-AS4 columns in series and a modified eluant, a satisfactory separation can be achieved and the particularly good separation of o- and p-methyl hippuric acids is especially noteworthy. Additionally, the elution sequence of these acids can be correlated with their pK_a values as can be seen from the data listed in Table 3-6.

The aliphatic carboxylic acids generally elute before the aromatic mono-acids and these elute before the aliphatic dicarboxylic acids.

The retention behavior of the aliphatic dicarboxylic acids is very similar to that of the bromide, nitrate, and sulphate anions. The capability of anion exchange chromatography for separating such compounds is demonstrated in Figure 3-31

TABLE 3-6: Elution sequence and pK_a values for some aromatic monocarboxylic acids.

ELUTION SEQUENCE	pK_a [15]
Mandelic acid	3.36
Hippuric acid	3.64
Benzoic acid	4.20

where a satisfactory separation of succinate, malonate, and maleate can be achieved using two HPIC-AS4 columns in series. In addition, the separation of the two stereoisomers maleic and fumaric acid is comparatively easy, since the latter has a retention time of approximately 20 minutes under the conditions used to obtain the chromatogram shown in Figure 3-31. In contrast to the aromatic monocarboxylic acids, the retention times of the dicarboxylic acids are inversely proportional to their pK_a values (see Table 3-7).

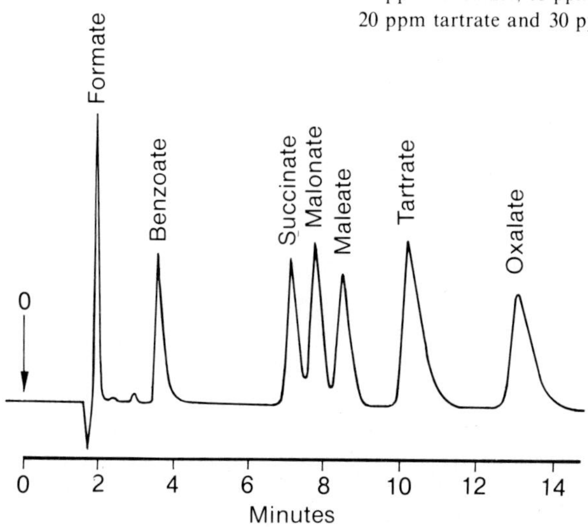

FIGURE 3-31.
The separation of some aliphatic dicarboxylic acids. **Column:** 2 HPIC-AS4. **Eluant:** 0.0028 mol/dm^3 $NaHCO_3$ and 0.0022 mol/dm^3 Na_2CO_3. **Flow rate:** 2 cm^3/min. **Detection:** Suppressed conductivity. **Sample:** $1 \cdot 10^{-4}$ mol/dm^3 sodium formate, 10 ppm benzoate, 20 ppm succinate, 10 ppm malonate, 30 ppm maleate, 20 ppm tartrate and 30 ppm oxalate.

This behavior can be explained in terms of the decrease in the charge stabilizing ($\lll$ denoted with a $<$ or $>$) of the methylene groups on going from succinic to oxalate.

TABLE 3-7: Elution sequence and pK_a values for some dicarboxylic acids.

ELUTION SEQUENCE	pK_a [15]
Succinic acid	4.21 (**a**), 5.64 (**b**)
Malonic acid	2.88 (**a**), 5.68 (**b**)
Oxalic acid	1.27 (**a**), 4.29 (**b**)

(**a**) First pK_a
(**b**) Second pK_a

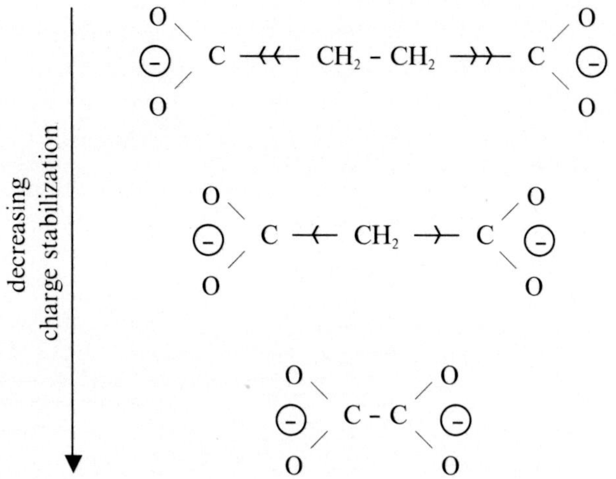

If the acid molecule also contains hydroxyl functions, such as is the case for tartaric acid, then the acid character of the molecule and its retention time increases. See Table 3-8.

Tricarboxylic acids, such as citric acid, exhibit unusually strong affinities for ion exchange resins and, therefore, cannot be eluted with low ionic strength carbonate/bicarbonate eluants.

TABLE 3-8: A comparison of the elution sequence and pK_a values of substituted and unsubstituted dicarboxylic acids.

ELUTION SEQUENCE	pK_a [15]
Succinic acid	4.21 (**a**), 5.64 (**b**)
Tartaric acid	3.04 (**a**), 4.37 (**b**)

(**a**) First pK_a
(**b**) Second pK_a

Such acids can be separated using an HPIC-AS5 column and by adding p-cyanophenol to the eluent. As an example, Figure 3-32 shows the chromatogram from a separation of 2,6-isonicotinic acid and thiocyanate which have very similar retention characteristics [16].

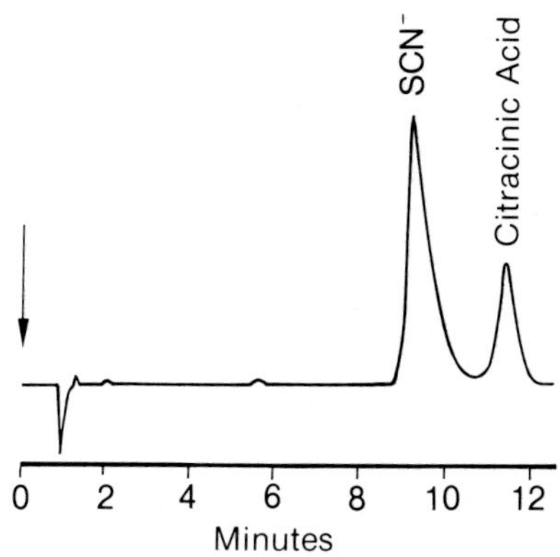

FIGURE 3-32.
The separation of citracinic acid. **Column:** HPIC-AS5. **Eluant:** 0.0028 mol/dm^3 $NaHCO_3$, 0.0022 mol/dm^3 Na_2CO_3 and 100 mg/dm^3 p-cyanophenol. **Flow rate:** 2 cm^3/min. **Detection:** Suppressed conductivity.

HPIC-AS1 to AS4 anion exchange columns are also suitable for the separation of inorganic and organic phosphates and sulphates. Generally, the organic ions elute prior to the inorganic ions. Typically, the longchain ($n_c > 4$) alkyl sulphates elute with the dead volume and can, thus, only be separated by ion pair chromatography. The shortchain alkyl sulphates and phosphates exhibit notable retentions. For the separation of monobutyl-, dibutyl-, and orthophosphate [17] (see Figure 3-33), a sodium carbonate/sodium hyroxide eluant was found to be optimal.

The high performance AS4 columns can also be used for separating biologically relevant phosphates. Among such are the glycerophosphates, which are involved in the metabolism of fatty acids to their esters. The latter are then stored in their glyceride forms. Additionally, the separation of individual RNA building blocks can also be achieved using a phosphate buffer eluant. Figure 3-34 shows a chromatogram of various nucleotides in which the adenine, cytosine, guanidine, and uracil bases are bound β-N-glycosidically to the 1′ site of the sugars. In this case, detection of the eluates is achieved by using their UV absorption at 254 nm.

The retention times are determined by the pH of the eluant since this controls the charge ratio of the pyrimidine and purine bases and also the degree of dissociation of the phosphate groups. AMP and CMP can also be eluted using standard conditions, but the selectivity is considerably affected by the basicity of the eluant. At a pH of 10.2, the free guanine and uracil bases are in their anionic forms and can both be detected in a single analysis (see Figure 3-35). The elution strength of carbonate is not enough to elute the corresponding di- and triphosphates and, in order to analyze these compounds, a concentrated phosphate buffer must be used.

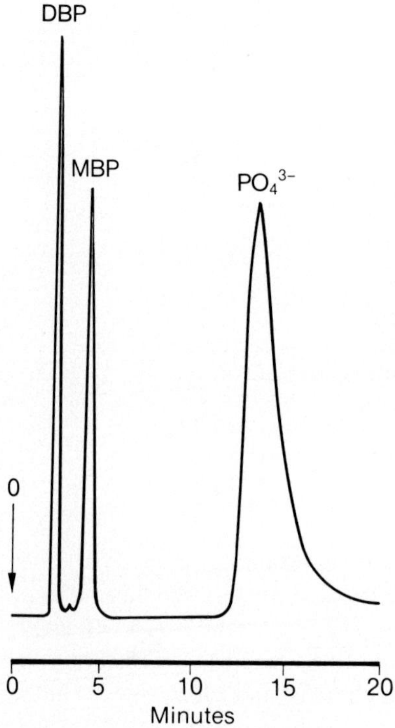

FIGURE 3-33.
The separation of organic and inorganic phosphates. **Column:** HPIC-AS1. **Eluant:** 0.003 mol/dm^3 Na_2CO_3 and 0.001 mol/dm^3 NaOH. **Flow rate:** 2.3 cm^3/min. **Detection:** Suppressed conductivity. **Sample:** 20 ppm DBP, 23 ppm MBP and 90 ppm PO_4^{3-} (100 μl).

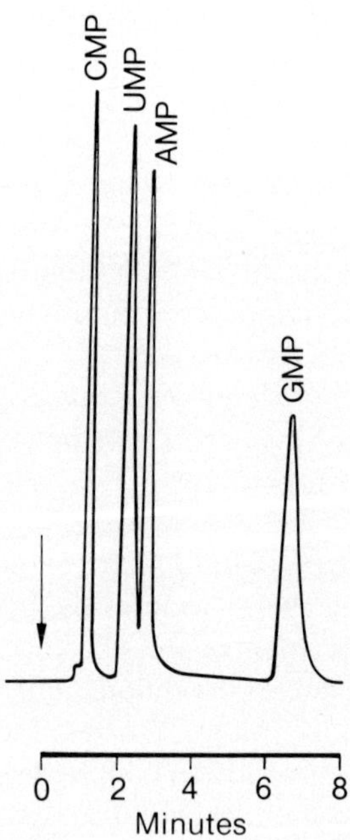

FIGURE 3-34.
The separation of biologically important phosphates. **Column:** HPIC-AS4. **Eluant:** 0.0025 mol/dm^3 NaH_2PO_4; pH adjusted to 3.35 with H_3PO_4. **Flow rate:** 1.5 cm^3/min. **Detection:** UV at 254 nm. **Sample:** 50 μl; 10 ppm CMP, UMP, AMP and GMP.

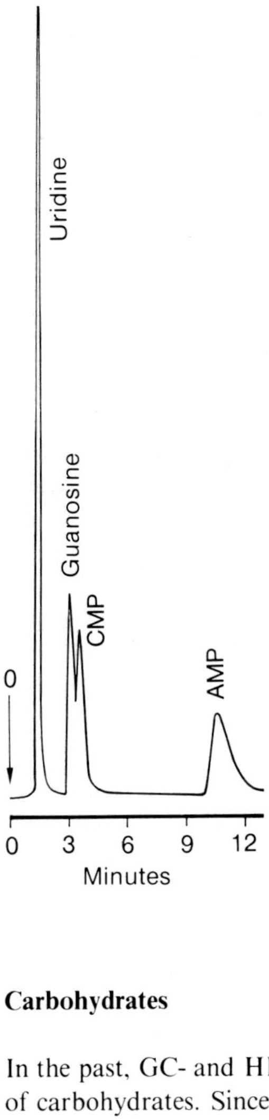

FIGURE 3-35.
The separation of purine and pyrimidine bases. **Column:** HPIC-AS4. **Eluant:** 0.0029 mol/dm^3 $NaHCO_3$ and 0.0023 mol/dm^3 Na_2CO_3. **Flow rate:** 2 cm^3/min. **Detection:** UV at 254 nm. **Sample:** 50 μl; 10 ppm uridine, guanosine, CMP and AMP.

Carbohydrates

In the past, GC- and HPLC-techniques have been developed for the separation of carbohydrates. Since the carbohydrates must first be derivatized to increase their volatility [18], gas chromatographic techniques are very time consuming. As an alternative, strong acid cation exchange resins, in their Ca-form, and a deionized water eluant have often been applied [19] to this analytical problem. However, due to a number of important disadvantages, this method is limited in its applicability:

- If the sample being analyzed contains organic acids, the metal ions of the resin can become complexed by these acids and, thus, eluted. This leads to the column requiring periodic regeneration.
- High capacity cation exchange columns have low selectivities for the oligosaccharides. The best results require low crosslinked resins. Unfortunately, the compressibility of such resins necessitate low flow rates and, thus, analysis times are very long.

- Optimal separations are obtained only with column temperatures of about 85° C. This leads to problems in the detection of the sugars due to changes in their refractive indices with temperature.
- The number of parameters for changing the retention times is limited, since only water or water/organic solvent mixtures can be used.

Good separations of mono- and disaccharides have been achieved with chemically bound amino groups on a silica gel support [20, 21], but the higher oligomers eluted as broad peaks. Furthermore, columns of this type have limited lifetimes if the samples being analyzed contain aldehydes or ketones. Foods and beverages often contain such compounds and these react with the amino groups to yield Schiff bases (imino compounds having the general structure: RR′C=NR″).

A fourth way of separating carbohydrates is via their boric acid complexes [22]. Although this technique can be made very selective, the kinetics of the complexation reactions are very slow, so that low flow rates have to be used and analysis times are very long.

An alternative to those techniques described above is to use anion exchange chromatography [23]. Since the pK_a values for sugars lie between 12 and 14 [24], a strongly alkaline eluant has to be used to convert them into their anionic forms and a strongly basic anion exchange resin in its hydroxide form is used for their separation. The major advantage of this technique over the methods described above is that the retention times and elution sequences can be varied by either the ionic strength of the eluant or the temperature of the column. The chromatogram from a separation of saccharides and alcoholic sugars using an HPIC-AS6 column is shown in Figure 3-36.

Raising the hydroxide concentration in the eluant leads to a reduction in the k′ values for all the solutes, allowing even the higher oligosaccharides to be analyzed using this technique. As an example of such a separation, Figure 3-37 shows a chromatogram obtained from a mixture of maltose oligomers (DP 2 to DP 10). By adding sodium acetate to the eluant, it is possible to separate even higher oligosaccharides.

When preparing the eluants for such separations, it is very important to ensure that they are free of carbonate. The carbonate ion is a stronger eluant than the hydroxyl ion for anion exchange chromatography and even traces of carbonate in the eluant can lead to a loss of resolution.

Anion exchange chromatography of carbohydrates leads to the following sequence of increasing affinity for the stationary phase: sugar alcohols < monosaccharides < disaccharides < oligosaccharides. Thus, during the separation of a mixture containing both di- and oligosaccharides, overlapping of peaks

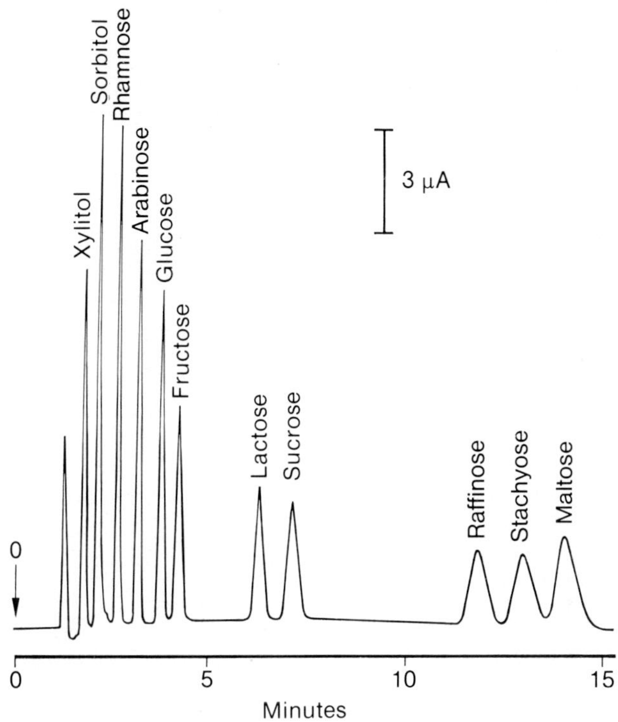

FIGURE 3-36.
The separation of some sugar alcohols and disaccharides. **Column:** HPIC-AS6. **Eluant:** 0.15 mol/dm^3 NaOH. T = 34°C. **Flow rate:** 1 cm^3/min. **Detection:** Pulsed amperometry at a gold working electrode. **Sample:** 25 ppm xylitol, 50 ppm sorbitol rhamnose, arabinose, glucose, fructose and lactose; 100 ppm sucrose, raffinose and stachyose; 150 ppm maltose.

may prove a problem. However, a change in the selectivity of the system can be achieved by altering the hydroxyl ion concentration. Figure 3-38 shows the dependence of retention times for some particular carbohydrates on the hydroxide concentration. It can be seen from Figure 3-38 that an optimum separation of the oligosaccharides is obtained at a NaOH concentration of 0.15 mol/dm^3.

A second parameter for controlling the retention times is the column temperature. Generally, a decrease in the retention time is observed with increasing temperature. The influence of the temperature on the retention time can be correlated with the size of the solute molecules and decreases in the sequence: oligosaccharides > disaccharides > sugar alcohols > monosaccharides. The dependence of retention time on column temperature is shown in Figure 3-39.

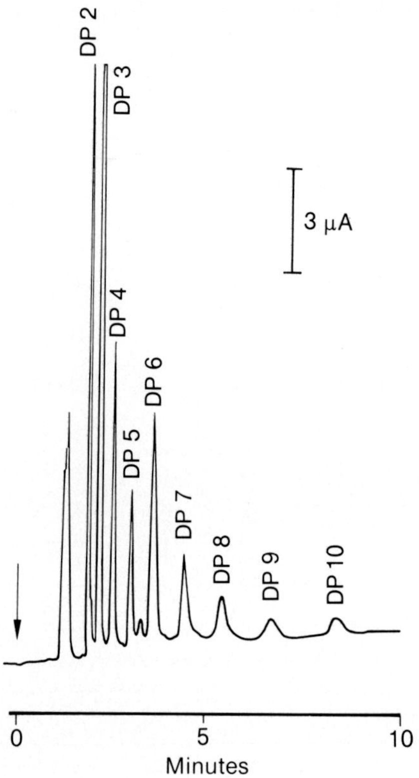

FIGURE 3-37.
The separation of maltose oligomers. **Column:** HPIC-AS6. **Eluant:** 0.2 mol/dm^3 NaOH and 0.2 mol/dm^3 sodium acetate. T = 34° C. **Flow rate:** 1 cm^3/min. **Detection:** Pulsed amperometry at a gold working electrode.

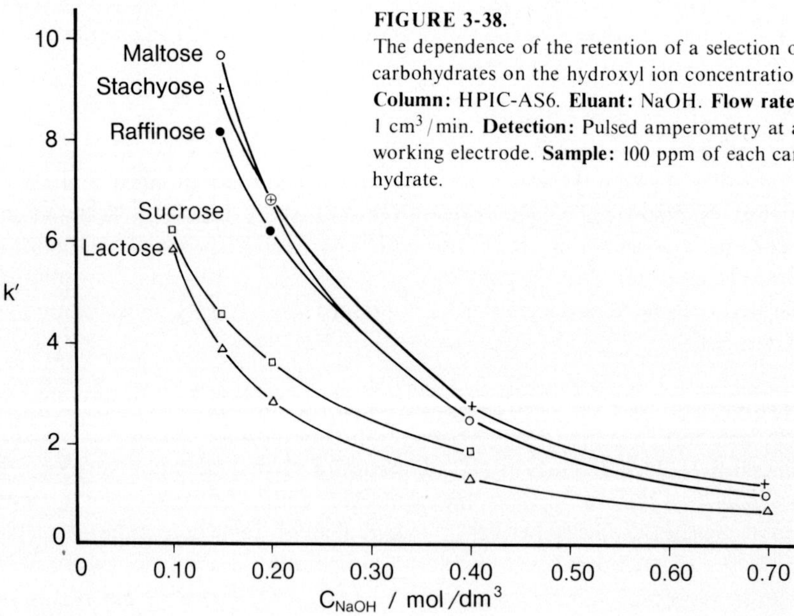

FIGURE 3-38.
The dependence of the retention of a selection of carbohydrates on the hydroxyl ion concentration. **Column:** HPIC-AS6. **Eluant:** NaOH. **Flow rate:** 1 cm^3/min. **Detection:** Pulsed amperometry at a gold working electrode. **Sample:** 100 ppm of each carbohydrate.

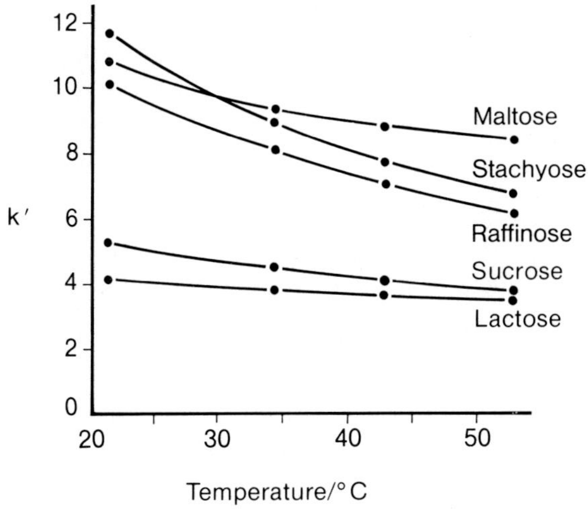

FIGURE 3-39.
Dependence of the retention of a selection of carbohydrates on the column temperature. **Column:** HPIC-AS6. **Eluant:** 0.15 mol/dm^3 NaOH. **Flow rate:** 1 cm^3/min. **Detection:** Pulsed amperometry at a gold working electrode. **Sample:** 100 ppm of each carbohydrate.

Particularly noteworthy is the reversal in elution sequence of maltose and stachyose with increasing column temperature. The column temperature cannot be raised indefinitely. Above 45° C the efficiency of the column starts to decrease and secondary signals and tailing effects are observed. Such effects can be ascribed to a Lobry-de-Bruyn-von-Ekenstein rearrangement. For example, glycerine aldehyde can transform, via an -enediol, into dihydroxy acetone. For these reasons, anion exchange columns are generally operated only between 20 and 45° C.

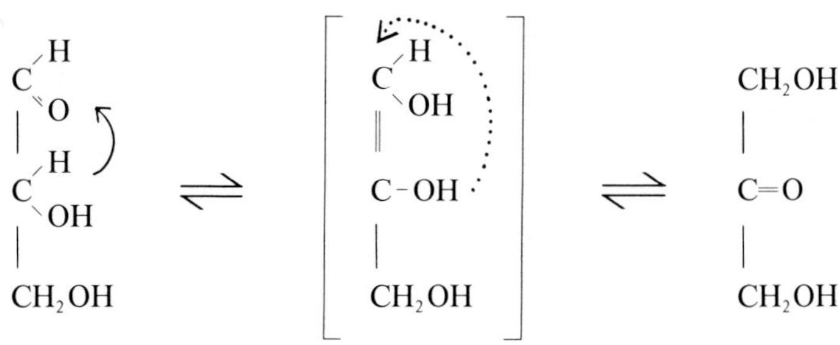

(48)

Glycerine aldehyde -enediol Dihydroxy acetone

Polyvalent Anions

Separation and determination of polyvalent anions by ion chromatography was, in the past, one of the most difficult analytical problems and was only possible for special cases. The reason for this difficulty lies in the chemistry of such compounds. Polyvalent anions exhibit a strong affinity for the stationary phases of exchange resins and, therefore, can only be eluted using very strong eluants. By appropriately changing the size of the latex particles and by altering the hydrophobicity of their functionality, a new type of anion exchange column with low exchange capacity has been developed (the HPIC-AS7 column). In conjunction with a nitric acid eluant, many polyvalent anions can be separated with these columns. Due to the high concentrations of nitric acid employed (0.03–0.07 mol/dm^3), conductivity detection ceases to be applicable. Thus, detection is achieved by postcolumn derivatization with iron (III) nitrate in acid solution and measurement of the UV absorption of the resulting complexes at 330 nm. The introduction of these columns and detection systems opened up a completely new analytical area for ion chromatography.

Among the many polyvalent anions which can be analyzed using this technique are the polyphosphonates. Figure 3-40 shows the chromatogram obtained from an analysis of a mixture containing Dequest polyphosphonates using nitric acid (0.03 mol/dm^3) as eluant. Such compounds have between 4 and 10 charges in their totally dissociated form and could not be eluted until the introduction of the AS7 columns. The use of a high acid concentration (pH 1.52) leads to a partial reduction in the degree of dissociation of these compounds and, thus, leads to a reduction in their retention times.

The reason for using nitric acid is that the nitrate ion is a strong eluant and a poor chelant for the Fe (III) used in the postcolumn detection (see Table 3-2). Not only the polyphosphonates, but also other complex forming species, can be determined by this form of ion chromatographic analysis. For example: citric acid, nitrilotriacetic acid (NTA), ethylene diamine tetraacetic acid (EDTA), and diethylene triamine pentaacetic acid (DTPA), can all be analyzed in this way and a representative chromatogram is shown in Figure 3-41.

Since all these separations involve materials that are capable of forming complexes, it is essential that the flow system be constructed from inert materials. Contamination by iron can lead to erroneous analytical results and, for this reason, only super-pure nitric acid should be used for the eluant.

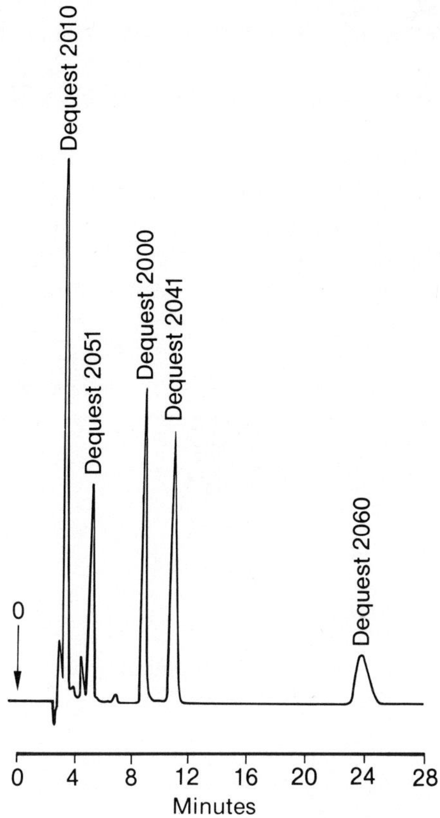

FIGURE 3-40.
The separation of Dequest type polyphosphonates. **Column:** HPIC-AS7. **Eluant:** 0.03 mol/dm^3 HNO_3. **Flow rate:** 0.5 cm^3/min. **Detection:** Photometry at 330 nm after reaction with $Fe(NO_3)_3$.

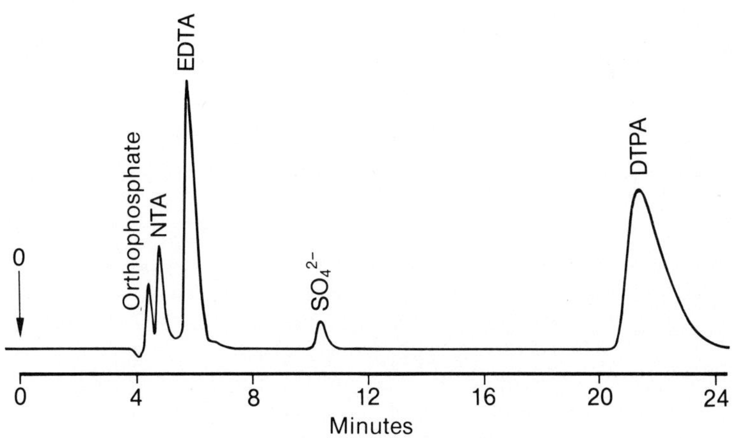

FIGURE 3-41.
The separation of complexes of NTA, EDTA and DTPA. **Column:** HPIC-AS7. **Eluant:** 0.05 mol/dm^3 HNO_3. **Flow rate:** 0.5 cm^3/min. **Detection:** Photometry at 330 nm after reaction with $Fe(NO_3)_3$.

3.3 CATION EXCHANGE CHROMATOGRAPHY

THE STRUCTURE OF THE STATIONARY PHASE

The stationary phase of cation exchange columns is also based on inert, surface-sulphonated, crosslinked polystyrene and corresponds to the basic structure used for an anion exchange column (see Section 3.2, "The Structure of the Support Material," earlier in this chapter). Figure 3-42 is a schematic of such a resin.

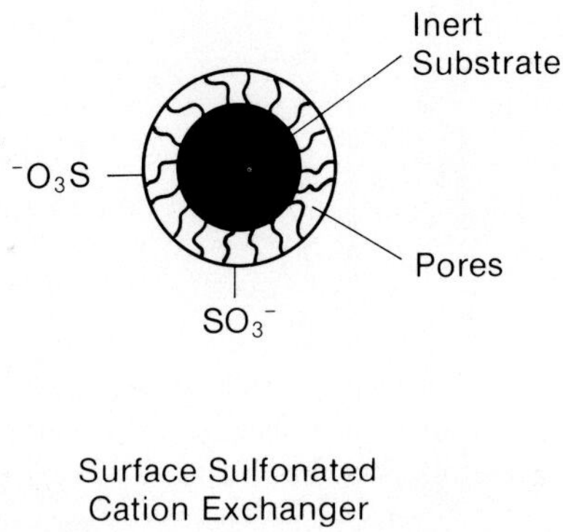

FIGURE 3-42.
Sketch of a cation exchange resin.

The exchange process for a cation M^+ occurring on such a resin can be formulated as follows:

$$\sim SO_3^-H^+ + M^+A^- \rightleftharpoons \sim SO_3^-M^+ + H^+A^- \qquad (49)$$

Since the core of a cation exchange resin is strongly hydrophobic, the diffusion into the resin of highly dissociated and hydrated species, such as Na^+, K^+, and Mg^{2+}, can be neglected. Consequently, the diffusion paths are short and high efficiencies—compared to conventional cation exchange materials—are achieved.

The cation exchange materials are also characterized by good pH stability (0 to 14). The problem of swelling in organic solvents, such as methanol, isopropanol, and acetonitrile (often encountered with conventional cation exchange materials), is less significant for the surface sulphonated resins.

SUMMARY OF THE VARIOUS CATION EXCHANGE RESINS AVAILABLE

At present, Dionex offers two different cation exchange resins: HPIC-CS1 and CS2. The exchange capacities of these resins are in the range of 0.005 to 0.1

mequiv./g. The CS1 columns, suitable for the analysis of the alkali and alkaline earth metals, contain particles having diameters of ca. 20 μm. For analyzing transition and heavy metals, the CS2 columns are employed which have a somewhat greater exchange capacity and particle diameters of ca. 15 μm. Both materials have a degree of crosslinking of ca. 2%.

Fritz and his coworkers [25] used the PS/DVB copolymer BN-X4, from Benson Co., for cation exchange chromatography without a suppressor system. This resin is only lightly sulphonated and has an exchange capacity of 0.02 mequiv./g. The spherical particles of this material, normally used for gel permeation chromatography, have diameters of ca. 20 μm. However, it is not possible to achieve a baseline separation of the alkali metals using this material.

The macroreticular XAD-1 polymers and the Vydac-SC materials have proved to be totally unsuitable for cationic exchange chromatography. With these materials, it is not even possible to separate completely the ammonium, sodium, and potassium cations.

ELUANTS FOR CATION EXCHANGE CHROMATOGRAPHY

A dilute (0.005 mol/dm^3) hydrochloric acid solution is often employed as an eluant for an analysis by cation exchange chromatography. The chloride is then exchanged for hydroxide in the suppressor system and the resultant product, water, has a comparatively low residual conductivity. Under such conditions, the alkali metals, ammonium, and aliphatic amines, with only a small number of C-atoms, can all be diluted. Divalent cations, such as the alkaline earth metals, cannot be eluted using dilute hydrochloric acid due to their increased affinity for the stationary phase. Simply increasing the concentration of the acid is not possible because:

- In a system *without* a suppressor unit, the high background conductivity would exclude the use of a conductivity detector.
- In a system *with* a suppressor unit, the background conductivity cannot be sufficiently reduced.
- The corrosiveness of hydrochloric acid solutions would become important.

Thus, in his introductory publication, Small [3] suggests the use of silver nitrate as eluant for the alkaline earth metals, since silver nitrate also has a strong affinity for the stationary phase. This eluant requires a suppressor of the chloride type so that the silver precipitates from the eluant in its insoluble chloride form. The use of this eluant, however, proved to be disadvantageous due to the broadening of the peaks and an increase in the backpressure of the suppressor column associated with the precipitation process. Thus, the use of silver nitrate eluants has been superseded by that of a mixture of m-phenylenediamine dihydrochloride and hydrochloric acid. Unfortunately, the amine hydrochloride tends to air oxidize (becoming violet) and, thus, the eluant mixture can only be stored for short periods.

For the elution of transition and heavy metal cations, organic acids which can form complexes are employed. Since these metals can be separated in their anionic and neutral forms, the nature and concentration of the organic acid(s) used depends on the type of separation technique involved. A summary of the various eluants which have been employed and the elution sequences that result is given in Table 3-9.

TABLE 3-9: Eluants for the analysis of the noble and heavy metals.

SEPARATION COLUMN	ELUANT	ELUTION SEQUENCE OF THE METALS TO BE ANALYZED
HPIC-CS2	0.01 mol/dm^3 oxalic acid, 0.0075 mol/dm^3 citric acid, pH 4.2 with LiOH.	Fe^{3+}, Cu^{2+}, Ni^{2+}, Zn^{2+}, Co^{2+}, Pb^{2+}, Fe^{2+}.
HPIC-CS2	0.04 mol/dm^3 tartaric acid, 0.012 mol/dm^3 citric acid, pH 4.3 with LiOH.	Ni^{2+}, Zn^{2+}, Co^{2+}, Pb^{2+}, Fe^{2+}, Cd^{2+}, Mn^{2+}.
HPIC-CS2	0.20 mol/dm^3 ammonium sulphate, 0.01 mol/dm^3 sulphuric acid.	Al^{3+}, Fe^{3+}.
HPIC-CS2	0.01 mol/dm^3 ammonium sulphate, 0.05 mol/dm^3 sulphuric acid.	UO_2^{2+}, Zn^{2+}.
HPIC-AS4	0.025 mol/dm^3 oxalic acid, pH 4.8 with NaOH.	Pb^{2+}, Cu^{2+}, Cd^{2+}, Mn^{2+}, Co^{2+}, Zn^{2+}, Ni^{2+}.

SUPPRESSORS FOR CATION EXCHANGE CHROMATOGRAPHY

In the past, the suppressors used for cation exchange chromatography were also columns containing an ion exchange resin. The resins used were strongly basic, microporous anion exchange resins such as Dowex 1, which is based on a PS/DVB copolymer. The resin contained quaternary ammonium bases and had a relatively high degree of crosslinking (ca. 10%). The good mechanical stability achieved by such a high degree of crosslinking also prevented swelling problems when organic eluants were used. Macroporous, strongly basic resins of the MP-1 type have not been accepted as an alternative suppressor material since the large surface area (200–800 m^2/g) of these materials led to undesirable adsorption effects with weak electrolytes.

The function of a suppressor has already been described in Section 3.2 of this chapter and does not require repeating here. The disadvantages of the conventional suppressor systems were also described and these apply equally to the conventional suppressors for cation exchange chromatography. Additionally, the suppressor columns for cation exchange chromatography are usually in their OH^- form and, thus, yield free bases of cations which are only weakly basic. For example, NH_4^+ is converted into NH_3 and, since this species is not ionic, it is not influenced by Donnan exclusion. The result is that such weak bases can be induced to interact with the stationary phase in the suppressor column which, in turn, leads to increased retention times. This leads to the fact that the retention times and peak heights of weak bases are dependent on the extent to which these conventional suppressor columns are exhausted. In order to minimize these effects, it is necessary to keep the pore volume small. However, in order to avoid adsorption effects, a low level of crosslinking is advantageous and this implies that large pore volumes are desirable. Thus, the degree of crosslinking in the resin is always a compromise, so that the effect of incomplete Donnan exclusion and adsorption effects are minimized.

As is generally known from HPLC, the size of the resin particles has an important influence on the width of the resulting peaks. Hence, peak broadening in a suppressor column is directly proportional to the size of the particles used. Furthermore, the dead volume of the suppressor column must be kept as low as possible in order to keep the peak broadening to a minimum. Thus, the best suppressor column should have a very small volume and contain an exchange resin with a very small particle diameter. These conditions are obviously totally contrary to normal analytical practice which requires a suppressor column that can be used for at least one complete working day without the exchange resin having to be regenerated. This condition can only be achieved with a suppressor column having a larger volume.

It is not surprising that, with all these disadvantages, the development of a membrane suppressor for cation exchange chromatography was attempted. The construction of such a suppressor is exactly the same as that used in anion exchange chromatography and described in Section 3.2, "Suppressor Systems in Anion Chromatography." Instead of the sulphate functions, the membrane contains quaternary ammonium bases in their OH^- form so that the counterions of the cations being analyzed can be exchanged for hydroxide ions. If the operating conditions are held constant, a dynamic equilibrium becomes established which retains a constant sensitivity with respect to weakly basic cations. With the introduction of the membrane suppressor and the concomitant reduction in dead volume, a reduction in peak broadening and an increase in the sensitivity of detection could be achieved.

Typical Operating Conditions

Conventional suppressors such as types CSC-1 or CSC-2 are regenerated using a solution of NaOH (0.5 mol/dm^3). The regenerant for the membrane suppressor is either 0.4 mol/dm^3 KOH solution or a 0.02 mol/dm^3 tetramethyl ammonium hydroxide solution. The choice of regenerating medium depends on the nature of the samples being analyzed. For alkali metals, a KOH solution is recommended; but if, in addition to alkali metal ions, ammonium is also present, it should be noted that the linear dynamic response range for detecting this ion—using an aqueous KOH regenerating medium—is very small. The reason for this is the equilibrium between the membrane suppressor product NH_4OH and the free base NH_3.

The flow rate of the regenerating medium should be between 2 and 3 ml/min and if an HCl eluent with a concentration of 0.005 mol/dm^3 is being used the background conductivity will then be about 7 μS/cm.

Aqueous KOH can also be used as regenerating medium when analyzing the alkaline earth metals, so long as a detection limit in the ppm range is sufficient. For detection in the ppb range, a tetramethyl ammonium hydroxide solution is recommended.

As already mentioned, eluants having greater ionic strengths will lead to higher residual conductivities, since increasing the concentration of the regenerating medium by a similar amount is not possible. At the higher concentrations, the relatively small potassium ion would overcome the Donnan exclusion forces and diffuse into the membrane core, leading to high background conductivities.

CATION EXCHANGE CHROMATOGRAPHY OF THE ALKALI METALS, THE ALKALINE EARTH METALS, AND AMINES

Monovalent cations, such as those from the alkali metals and smaller aliphatic amines, can be separated using dilute hydrochloric acid as eluant and an example of such a separation is shown in Figure 3-43.

The elution sequence for the alkali metals (see Figure 3-43) results from the affinity of the cations for the stationary phase becoming stronger with increasing ionic radius. The advantages of ion chromatography, as compared to traditional methods, for the analysis of the alkali metals lies not in any improvement in sensitivity but in that several metals can be analyzed simultaneously. The key ion in the mixture is, without doubt, the ammonium ion (in Figure 3-43, the peak between those of sodium and potassium). The sensitive analysis of this ion in such a mixture is extremely difficult using other techniques.

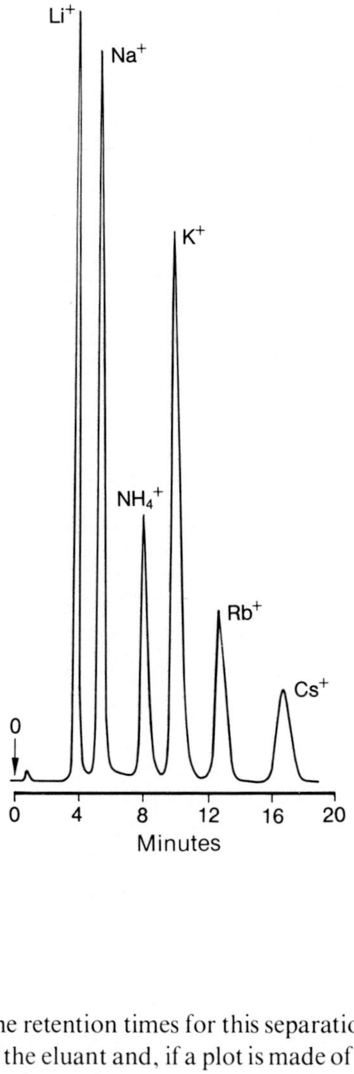

FIGURE 3-43.
The separation of alkali metals on a surface-sulphonated cation exchange resin. **Column:** HPIC-CS1. **Eluant:** 0.005 mol/dm^3 HCl. **Flow rate:** 2.3 cm^3/min. **Detection:** Suppressed conductivity. **Sample:** 50 μl contain-in 5 ppm Li^+ and Na^+; 10 ppm NH_4^+ and K^+; 20 ppm Rb^+ and 30 ppm Cs^+.

The retention times for this separation are determined solely by the concentration of the eluant and, if a plot is made of ln k′ as a function of ionic strength, a straight line is obtained as can be seen in Figure 3-44.

As well as the alkali metal cations, it is also possible to analyze aliphatic amines — provided these have only a few C-atoms — using the experimental conditions used to produce the chromatogram shown in Figure 3-43. The most important amines that can be separated and eluted as their cations include: the mono-, di- and trimethyl amines as well as the corresponding ethyl derivatives. The disadvantages of the method are the long retention times (between 10 and 30 minutes) and possible interference between peaks if cesium and/or rubidium are present in the sample.

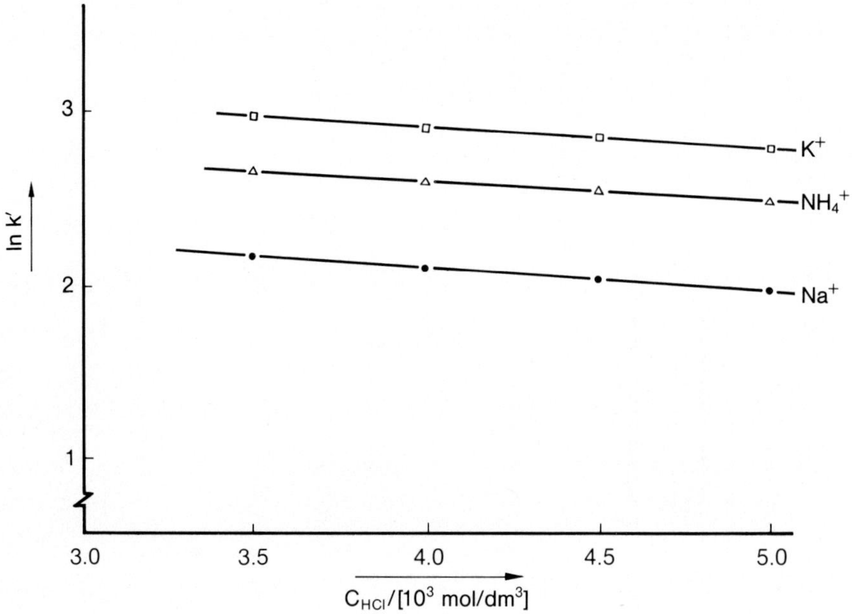

FIGURE 3-44.
The dependence of the retention of monovalent cations on the ionic strength of the eluant with Na^+, NH_4^+ and K^+ as examples. **Column:** HPIC-CS1. **Eluant:** HCl. **Detection:** Suppressed conductivity.

Figure 3-45 shows the chromatogram obtained from an analysis of a mixture of the alkaline earth metals using a mixture of m-phenylene diamine dihydrochloride and aqueous HCl as eluant. As can be seen, the ions elute in the sequence: $Mg^{2+} < Ca^{2+} < Sr^{2+} < Ba^{2+}$.

Because of the strength of this eluant, the alkali metals (also present in the sample mixture) all elute with the dead volume. The "fronting," which both the Mg^{2+} and the Ca^{2+} peaks exhibit in Figure 3-45, has already been discussed.

THE SEPARATION OF TRANSITION AND HEAVY METALS

As early as 1939, Samuelson [26] reported a separation of transition and heavy metals using ion exchange chromatography. The resin used in this early work was of a straightforward ion exchange type. The detection system used for quantifying the results involved manual spectrophotometric analysis of the individual fractions.

It wasn't until the introduction of high performance cation and anion exchange resins that the analysis of transition and heavy metals was significantly improved.

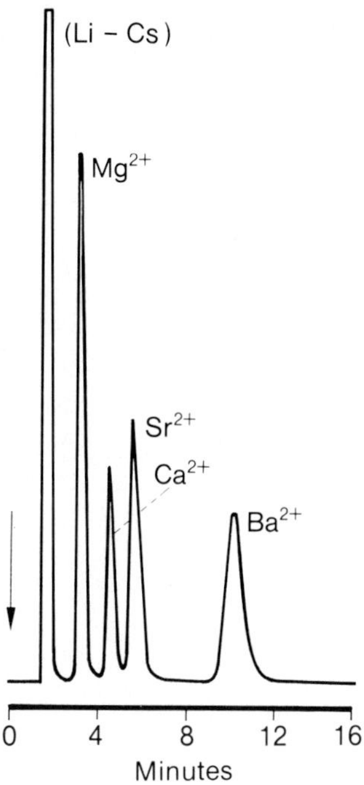

FIGURE 3-45.
A separation of the alkaline earth metals on a surface-sulphonated cation exchange resin. **Column:** HPIC-CS1. **Eluant:** 0.002 mol/dm^3 m-phenylenediamine dihydrochloride and 0.002 mol/dm^3 HCl. **Flow rate:** 2.3 cm^3/min. **Detection:** Suppressed conductivity. **Sample:** 3 ppm Mg^{2+} and Ca^{2+}, 10 ppm Sr^{2+} and 25 ppm Ba^{2+}.

The separation of transition metals on a *cation exchange* column requires complexation of the metal by the eluant in order to reduce its charge density. Thus, monovalent cations, such as Na^+ or H^+, are not suitable as eluants. Since the selectivity coefficients for transition metal complexes having the same charge are essentially the same, changes in selectivity with respect to a cation exchange column can only be achieved via the formation of neutral or anionic complexes during the separation by employing appropriate complexing agents. Typical complexing agents are the weak organic acids which preferentially form anionic complexes with the transition metals; i.e., citric acid, oxalic acid, and tartaric acid. The results of separations using these acids are considerably better than those obtained with compounds that form cationic complexes, such as ammonium hydroxide and ethylene diamine. An optimum separation can be obtained if a mixture of two different complex forming acids are used as eluant. In order to simplify the appreciation of the kinetics of these systems, the following discussion considers only a single ligand.

Assuming that the solution being considered contains a low concentration of a multivalent transition metal ion, M^{n+}, and a monovalent cation, A^+ (Na^+ or H^+), whose concentration is considerably greater than that of M^{n+}, then an equilibrium will be established if this solution comes into contact with a cation exchange material:

$$M^{n+} + nRA \rightleftharpoons R_nM + nA^+ \quad (50)$$

where R = resin.

From the mass balance equation, the equilibrium constant K_{MA} can be formulated as:

$$K_{MA} = \frac{[R_nM][A^+]^n}{[RA]^n[M^{n+}]} \quad (51)$$

If only ion exchange processes are operative, then the distribution coefficient D_M is given simply by the ratio of the concentrations in the mobile and stationary phases (when the system is at equilibrium) of the ion being investigated. Thus:

$$D_M = \frac{[R_nM]}{[M^{n+}]} = K_{MA}\frac{[RA]^n}{[A^+]^n} \quad (52)$$

Since the metal ion concentration is negligibly small compared with that of the eluent:

$$[M^{n+}] << [A^+] \text{ and } [R_nM] << [RA] \quad (53)$$

From the above treatment, it can be seen that K_{MA} is a constant for a given eluant concentration. Since RA is, to a reasonable approximation, constant and denoting the exchange capacity of the resin by C, it is possible to derive the distribution coefficient from:

$$D_M \approx \frac{K_{MA} \cdot C^n}{[A^+]^n} \quad (54)$$

which can be rearranged to a logarithmic form

$$\log D_M \approx \log K_{MA} + n \log C - n \log [A^+] \quad (55)$$

However, it is more usual to measure the capacity factor k′ (Equation 6, Chapter 2) rather than the distribution coefficient D_M.

From Equation 55, it can be seen that the distribution coefficient is directly proportional to the concentration of the eluant A^+. A plot of log D_M vs. log A^+ thus gives a straight line whose slope is equal to the number of charges on the ion n.

Hence, the selectivity of a separation of metal ions having similar charges is determined solely by their K_M values. The concentration of the cation A^+ in the eluent makes only a small contribution to the selectivity, since the variation in the logarithm of the distribution coefficients with eluant concentrations is constant. For these reasons, satisfactory separations can only be achieved if the selectivity coefficients of two transition metal ions, M_1^{n+} and M_2^{n+}, are significantly different. Only if the ionic charge is different is the selectivity of the column dependent on the eluant concentration.

If the eluant contains complex forming molecules, such as citric or oxalic acid, then the equation for the distribution coefficient of the metal ion M^{n+} must be modified to include the competitive equilibrium associated with the formation of the corresponding complexes. The coefficient of complex formation can be formulated in the following way:

$$a_{M(L)} = 1 + [L] \cdot B_{ML} + [L]^2 \cdot B_{ML_2} + \ldots . \qquad (56)$$

$a_{M(L)}$: Coefficient of complex formation.

$[L]$: Concentration of the anionic ligand.

B_{ML}: Formation constant of the complex formed from:

$$M^{n+} = L^{x-} \rightleftharpoons ML^{n-x}$$

B_{ML_2}: Formation constant for the equilibrium:

$$M^{n+} + 2\,L^{x-} \rightleftharpoons ML_2{}^{n-2x}$$

If the complexing form of the ligand is a strong conjugated base, then the protonation of the uncomplexed ligand must be taken into account when calculating a_{ML}. Since the concentration of the complex forming ligand is much greater than that of the metal ion under normal chromatographic conditions, the complexation equilibrium is hardly influenced by the partial protonation of the ligand. This allows the distribution coefficient of a metal ion in the presence of a complex forming ligand to be formulated as:

$$\log D_M = \log K_{MA} - \log a_{M(L)} + n \log C - n \log [A^+] \qquad (57)$$

The optimum separation of a particular metal ion is achieved by adjusting the pH of the eluant. If weak organic acids are employed as ligands, then a reduction in the pH of the eluant reduces the effective concentration of the ligand and leads to longer retention times. If, alternatively, the effective concentration of the ligand is increased by increasing the concentration of the organic acid in the eluant, then the concentration of the counterion (Na^+ or H^+) will also increase. This would lead to a faster displacement of the metal ions from the exchange sites on the resin which would result in a loss of separation capability.

For cation exchange columns having low capacities, the exchange capacity is dependent on the nature of the counterion. Therefore, it is recommended that the pH of the eluent be adjusted by using LiOH since Li^+ has less affinity for the stationary phase than Na^+.

The separation of transition and heavy metals with an anion exchange system also utilizes the ability of these metals to form anionic complexes.

Separations of this type rely on the significant stability differences of the complexes formed from divalent anionic complexing agents. Superb separations have been achieved [27] with pellicular anion exchange materials and an oxalic acid eluant. The coefficient of the distribution of a metal ion, between the mobile and stationary phases in the presence of a divalent anionic ligand, is given by:

$$D_M = \frac{[R_{n-x} \cdot ML_x]}{[M^{n+}]} = K_M \cdot \Phi_x \cdot \left(\frac{[RL]}{[L^-]}\right)^{n-x} \qquad (58)$$

Where K_M is the selectivity coefficient for the anions, ML^{n-x} and L^- and can be described by:

$$K_m = \frac{[R_{n-x} \cdot ML_x][L^-]^{n-x}}{[RL]^{n-x}[ML_x^{\ n-x}]} \qquad (59)$$

Where Φ_x is the mole fraction of the complexes $ML_x^{\ n-x}$ in solution, this can be calculated from:

$$\Phi_x = \frac{[L^-]\,\beta_{ML_x}}{1 + [L^-]\,\beta_{ML} + [L^-]^2 \cdot \beta_{ML_2} + \ldots} \qquad (60)$$

However, under normal chromatographic conditions

$$[M^{n+}] << [L^-]$$

so that the free ligand concentration L^- can be substituted by the total ligand concentration L_t. If, in addition, [RL] is then substituted by the exchange capacity C, one obtains:

$$D_M = K_M \cdot \Phi_x \cdot C^{n-x} \cdot [L^-]^{n-x} \qquad (61)$$

which can be rewritten in a logarithmic form:

$$\log D_M = \log K_M + \log \Phi_x + (n-x)\log C + (n-x)\log [L^-] \qquad (62)$$

In the cation exchange chromatography of transition metals, an increase in the concentration of the complexing ligand leads to a loss of resolution. In contrast, an optimal separation can be obtained by anion exchange chromatography using a concentration of ligand which leads to the formation of neutral complexes. If the concentration of ligand is then increased, the formation of anionic complexes increases and this leads to longer retention times. This effect is partially offset since the free anionic ligand is also responsible for the elution of the anionic complexes.

The choice of ligand for a particular separation of transition metals using a cation exchange resin should take into account:

- Metal ion and ligand must form neutral or anionic complexes.
- Large differences in the formation constants of the complexes of the various metals lead to increased selectivity.

- The transition metal-ligand complex should be thermodynamically stable but kinetically labile (i.e., the formation of the complex must be thermodynamically possible but reversible under the chromatographic conditions used). This requirement is demonstrated using two examples:
  a) If the transition metal complex has a large formation constant but is kinetically labile, the retention mechanism will involve only ion exchange processes. Long retention times will result.
  b) If the formation constant for the complex is small and the complex kinetically labile, strong tailing effects will be observed.
- If detection of the eluting species is achieved by postcolumn derivatization with an appropriate indicator, the equilibrium constant for the following reaction should be large and the indicator complex (Me In) kinetically stable.

$$\text{Me L}_x + \text{In} \rightleftharpoons \text{Me In} + x\text{L} \tag{63}$$

Me: Metal.
L: Ligand.
In: Indicator.

Similar considerations must be taken into account when separating transition metals using anion exchange columns. In this case, the size of the ligand should be relatively small so that the resulting complex has roughly the same size as the hydrated metal ion.

Transition metals can also be separated and analyzed using ion pair chromatography on macroporous PS/DVB resins or chemically bound phases on a silica gel support [28]. The mobile phase contains complexing agents and an appropriate ion pair reagent. For the analysis of cations, a suitable ion pair reagent is octane sulphonc acid. With such a system, the elution sequence for the ions Cu^{2+}, Ni^{2+}, Zn^{2+}, and Co^{2+}, is the same as that obtained using a CS-2 column. From this result, one can conclude that the ion exchange process dominates over the ion pair formation process. However, initial experiments suggest that the quality of the separation of cations using an MPIC-phase is not as good as that obtained using surface sulphonated cation exchange resins.

The technique of separating heavy metals, using anion exchange resins of type HPIC-AS4, can also be adapted to ion pair chromatography. With oxalic acid as complexing agent and tetrabutyl ammonium hydroxide as ion pair reagent, one obtains an elution sequence which is the reverse of that obtained by cation analysis. The separation of anionic metal complexes on HPIC-phase occurs solely via anion exchange processes. In contrast, using latex anion exchange materials, the separation process involves (due to the way in which these materials are constructed) both cation and anion exchanges.

Separations by Cation Exchange

Figure 3-46 shows the chromatogram obtained using an HPIC-CS2 cation exchange column for the separation of a mixture of heavy metals. Although satisfactory separation can be achieved using only oxalic acid as the complexing agent, increased selectivity is obtained by adding citric acid to the eluant as a second complexing agent. From Figure 3-46, it can be seen that Fe^{3+} elutes with the dead volume peak. Complete exclusion of Fe^{3+} can be attributed to the formation of the thermodynamically stable $Fe(Ox)_3^{3-}$ complex. Due to the unusual stability of this species (log K = 18.5) [29], it does not interact with the stationary phase and, thus, elutes with the dead volume peak. Note also the poor separation of Fe^{2+} and Cu^{2+}. For the best separation of these two ions, it is important that the pH of the sample be adjusted to 1.5. In Figure 3-47, the capacity factor k′ is plotted as a function of pH for the Fe^{2+} ion. At higher pH,

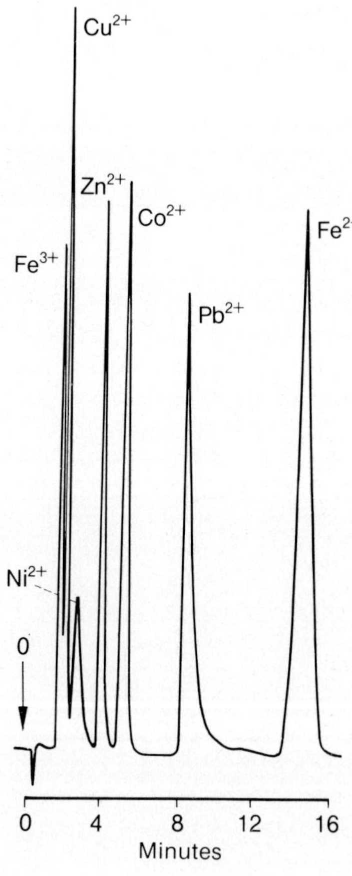

FIGURE 3-46.
The separation of transition and heavy metals. **Column:** HPIC-CS2. **Eluant:** 0.01 mol/dm^3 oxalic acid and 0.0075 mol/dm^3 citric acid; pH adjusted to 4.2 with LiOH. **Flow rate:** 1 cm^3/min. **Detection:** Photometry at 520 nm after reaction with PAR. **Sample:** 50 μl containing 5 ppm Fe^{2+} and Fe^{3+}, 0.5 ppm Cu^{2+}, Ni^{2+} and Zn^{2+}, 1 ppm Co^{2+} and 10 ppm Pb^{2+}.

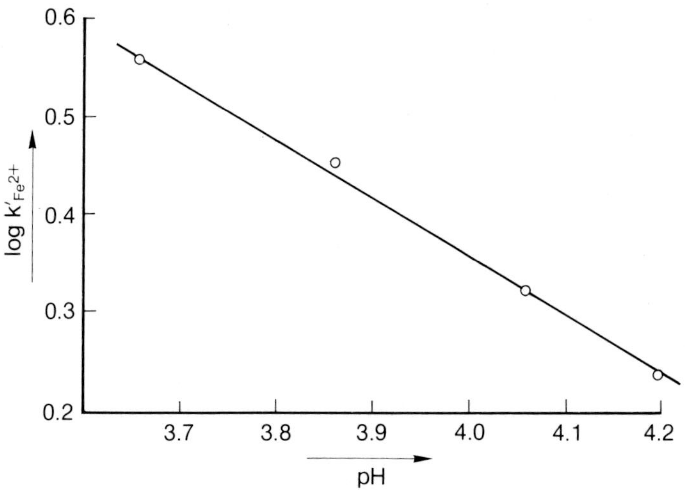

FIGURE 3-47.
The dependence of the retention on the pH of the eluant Fe^{2+} as example. **Column:** HPIC-CS2. **Eluant:** 0.01 mol/dm^3 oxalic acid and 0.018 mol/dm^3 tartaric acid. **Detection:** Photometry at 520 nm after reaction with PAR.

the concentration of the complexing form of the organic acid increases and this leads to a greater concentration of the complex which, in turn, leads to less frequent ion exchange and a reduction in k'.

Cadmium and manganese are strongly adsorbed by the stationary phase under the experimental conditions used to obtain the chromatogram shown in Figure 3-46, but an optimal separation of these metals can be achieved by using a mixed citric and tartaric acid eluant (see Figure 3-48). Resolution and analysis time for both these separations are determined by the pH of the eluant.

Analysis of the uranyl cation UO_2^{2+} is also possible with the aid of a cation exchange column. Since this ion interacts only weakly with the stationary phase, a buffer solution containing ammonium sulphate and sulphuric acid is used as eluant. Figure 3-49 shows the chromatogram obtained from a separation of UO_2^{2+} and Zn^{2+}. Buffer solutions can also be used for eluting aluminium which can then be detected by postcolumn derivatization with Tiron (4,5-dihydroxyl-1, 3-benzene disulfonic acid-disodium salt). See Figure 3-50. In an acid sulphate medium the extent to which the aluminum ions are in the $AlSO_4^+$ form depends on the proton and sulphate ion concentrations. In this case, the retention times are determined by the ammonium sulphate concentration.

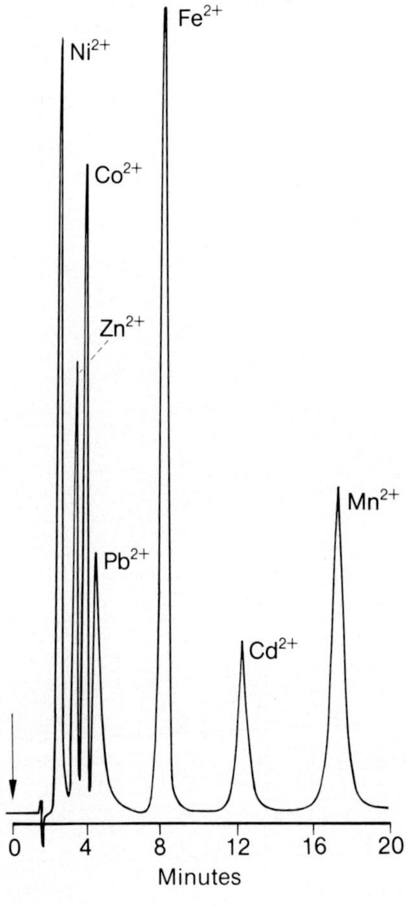

FIGURE 3-48.
The separation of cadmium and manganese. **Column:** HPIC-CS2. **Eluant:** 0.04 mol/dm^3 tartaric acid and 0.012 mol/dm^3 citric acid; pH adjusted with LiOH to 4.3. **Flow rate:** 1 cm^3/min. **Detection:** Photometry at 520 nm after reaction with PAR. **Sample:** 50 μl containing 1 ppm Ni^{2+}, 0.5 ppm Zn^{2+} and Co^{2+}, 5 ppm Pb^{2+}, 3 ppm Fe^{2+} and Cd^{2+} and 2 ppm Mn^{2+}.

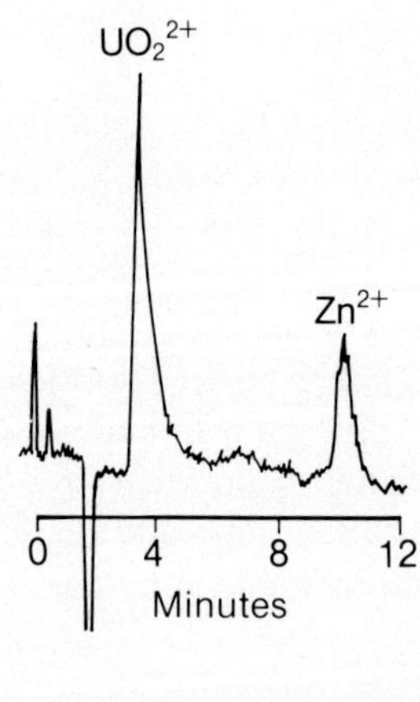

FIGURE 3-49.
Separating trace amounts of the uranyl cation. **Column:** HPIC-CS2. **Eluant:** 0.01 mol/dm^3 $(NH_4)_2SO_4$ and 0.05 mol/dm^3 H_2SO_4. **Flow rate:** 1 cm^3/min. **Detection:** Photometry at 520 nm after reaction with PAR. **Sample:** 50 μl containing 10 ppb UO_2^{2+} (and 5 ppb Zn^{2+} as impurity).

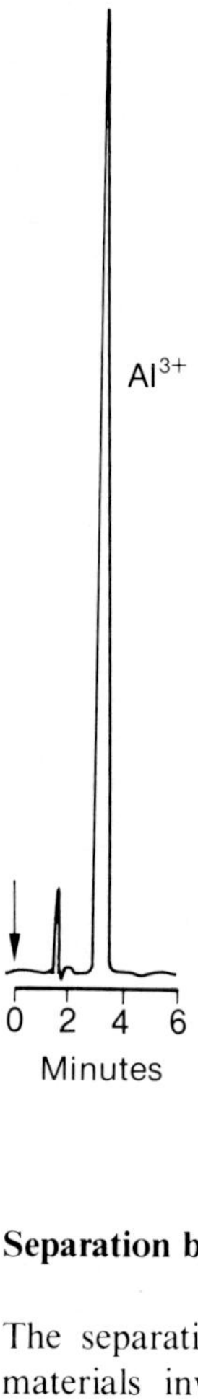

FIGURE 3-50.
The separation of aluminium. **Column:** HPIC-CS2. **Eluant:** 0.2 mol/dm^3 $(NH_4)_2SO_4$ and 0.1 mol/dm^3 H_2SO_4. **Flow rate:** 1 cm^3/min. **Detection:** Photometry at 313 nm after reaction with Tiron. **Sample:** 50 μl containing 5 pm Al^{3+}.

Separation by Anion Exchange

The separation of transition and heavy metals using latex anion exchange materials involves both cation and anion exchange processes. The cation exchange capacities of these materials is a result of their surface sulphonation, and the ratio between the anion and cation exchange capacities controls the unusually high selectivity of this technique. Figure 3-51 shows the chromatogram obtained from a separation of several heavy metals using an HPIC-AS4 column and oxalic acid as eluant. Noteworthy is the reversed elution sequence for the Pb^{2+}, Co^{2+}, Zn^{2+}, and Ni^{2+} ions as compared to that obtained using a cation exchange technique. For the reasons discussed above, the Mn^{2+} and Cu^{2+} ions do not follow the same trend.

The influence of the pH of the eluant on the retention times has been studied using Cu^{2+}, Co^{2+}, and Zn^{2+}. Maximum k′ values are obtained by forming weak anionic complexes of these ions in the eluant. Figure 3-52 shows that the k′ values for Co^{2+} and Zn^{2+} reach a maximum at a pH of 3.5. The value of k′, together with the selectivity, indicate that an anion mechanism is operating. At higher pH,

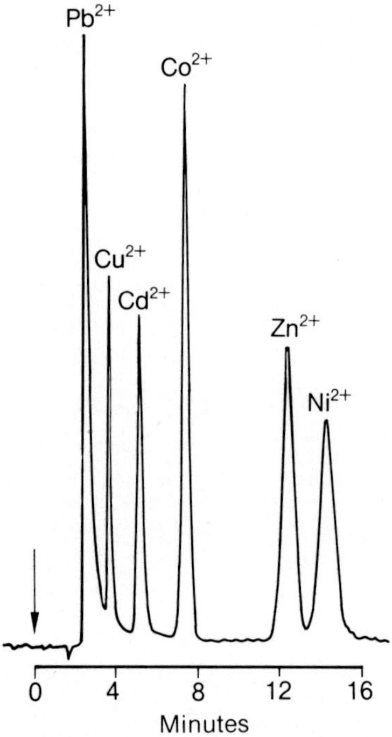

FIGURE 3-51.
Separating transition and heavy metals using anion exchange chromatography. **Column:** HPIC-AS4. **Eluant:** 0.025 mol/dm^3 oxalic acid adjusted to pH 4.8 with NaOH. **Flow rate:** 1 cm^3/min. **Detection:** Photometry after reaction with PAR. **Sample:** 50 μl containing 3 ppm Pb^{2+} and Cd^{2+}, 0.2 ppm Cu^{2+}, 0.5 ppm Co^{2+} and Ni^{2+} and 1 ppm Zn^{2+}.

the k′ values indicate that a cation exchange mechanism is also operating. For Cu^{2+}, k′ does not reach a maximum value within the pH range studied. Hence, a simple cation exchange mechanism must be assumed for the separation of this ion.

Although the separation of metal ions on an HPIC-AS4 column involves both anion and cation exchange processes, the extent to which the individual processes operate is different for each metal ion. Thus, if stable anionic complexes can be formed, an anion exchange mechanism will dominate. Whereas for metals which don't form stable oxalate complexes, a cation exchange mechanism predominates.

Under the conditions used to obtain the chromatogram shown in Figure 3-51, it is not possible to elute Fe^{3+} and Fe^{2+}. The strong affinity of Fe^{3+} for the stationary phase can be attributed to the formation of the stable oxalate complex $Fe(Ox)_3{}^{3-}$. The divalent oxalate anion is not strong enough to elute this trivalent species. This reasoning cannot be applied to the Fe^{2+} ion and, although it is presumed that this ion is oxidized on the column to Fe^{3+}, there is no direct experimental corroboration of this to date.

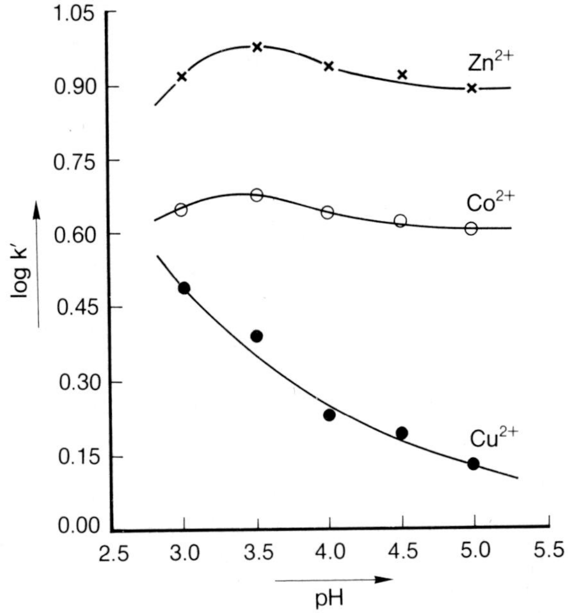

FIGURE 3-52.
Dependence of the retention of Cu^{2+}, Co^{2+} and Zn^{2+} on the pH of the eluant using anion exchange columns. **Column:** HIPC-AS4. **Eluant:** 0.25 mol/cm^3 oxalic acid. **Flow rate:** 1 cm^3/min. **Detection:** Photometry at 520 nm after reaction with PAR.

ANALYSIS OF POLYAMINES

As well as alkali, alkaline earth, and transition and heavy metals, it is also possible to analyze polyamines such as cadaverine, spermine, etc., using cation exchange columns. Such compounds can be detected via the fluorescence of the products of their reaction with a o-phthalaldehyde. In the past, the separation of polyamines was a difficult analytical problem since these compounds exhibit an extreme affinity for the stationary phase of traditional ion exchange resins. Thus, for such analyses, buffer solutions with high ionic strengths had to be used and these had a deleterious effect on the useful lifetime of the chromatographic columns. Compression of the resins led to dead volume at the entrance to the column, increasing backpressure and drastic loss of column efficiency. After the introduction of the surface-sulphonated, pellicular cation exchange materials, it was possible to halve the ionic strength of the eluant—leading to extended column life, while at the same time, increasing column efficiency and reducing analysis times. Figure 3-53 shows a standard chromatogram from a separation of putresine, cadaverine, spermidine, and spermine—all in nmol concentrations. The eluant composition necessary for such a separation is given in Table 3-10.

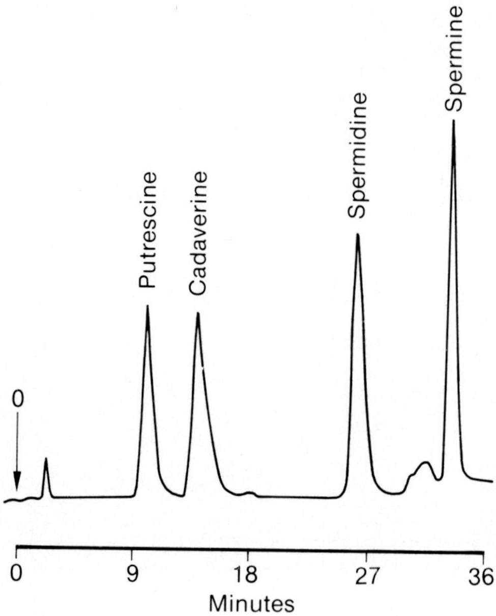

FIGURE 3-53.
The separation of polyamines. **Column:** HPIC-CS1. **Eluant:** Citrate buffer (see text for details). **Flow rate:** 0.6 cm^3/min. **Detection:** Fluorescence after reaction with o-phthalaldehyde. **Sample:** 20 μl containing 1 nmol of each of the following polyamines: putresine, cadaverine, spermine and spermidine.

TABLE 3-10: Eluant composition for the separation of polyamines.

BUFFER	pH	CATION CONCENTRATION	TRISODIUM CITRATE	SODIUM CHLORIDE
A	5.80	0.3 mol/dm^3	0.98 g/dm^3	16.94 g/dm^3
B	5.55	0.7 mol/dm^3	1.97 g/dm^3	42.63 g/dm^3
C	5.55	1.5 mol/dm^3	7.84 g/dm^3	83.00 g/dm^3
Regenerant	—	0.2 mol/dm^3	—	5.84 g/dm^3

BUFFER	PHENOL	EDTA	NaOH
A	1.0 g/dm3	—	—
B	1.0 g/dm^3	—	—
C	1.0 g/dm^3	—	—
Regenerant	—	0.25 g/dm^3	4.01 g/dm^3

LITERATURE:

1. Exodus, **15,** 24 and 25.
2. W. Riemon and H.F. Walton, "Ion Exchange in Analytical Chemistry," Pergamon Press 1970.
3. H. Small, T.S. Stevens and W.C. Baumann, Anal. Chem. **47,** (1975).
4. C. Horvath, "Pellicular Ion Exchangers. Bonded Stationary Phases in Chromatography," Edited by E. Grushka, Ann Arbor Science 1974.
5. J.S. Fritz and J.N. Story, J. of Chromatogr. **90,** 267 (1974).
6. D.R. Jenke and G.K. Pagenkopf, Anal. Chem. **55,** 1168 (1983).

7. G. Ackermann et al, "Elektrolytgleichgewichte und Elektrochemie," VEB Verlag für Grundstoffindustrie, Leipzig 1974.
8. Holleman-Wiberg, "Lehrbuch der anorganischen Chemie," 71st–80th edition, Walter de Gruyter & Co. 1971.
9. H.J. Möckel, Lecture "Instrumentelle Analytik I," TU Berlin, 1974–1984.
10. H.J. Möckel, J. of Chromatogr. 1984, in press.
11. H.J. Möckel, T. Freyholt, J. Weiβ and I. Molnar, "The HPLC of Divalent Sulfur" in I. Molnar (Editor): "Practical Aspects of Modern HPLC," Walter de Gruyter and Co., Berlin, New York 1982.
12. C. Eon, Anal. Chem. **47,** 1871 (1975).
13. C. Eon and G. Guiochon, J. Colloid. and Interface Sci. **45,** 521 (1973).
14. J. Weiβ and G. Storck, unpublished results.
15. P.D. Perrin and B. Dempsey, "Buffers for pH and Metal Ion Control," Chapman and Hall, London 1974.
16. Dionex GmbH Weiterstadt, Application Note Nr. 1/84/3.
17. Dionex Corp. Sunnyvale, Application Note Nr. 16.
18. P.R. Knapp, "Handbook of Analytical Derivatisation Reactions," Wiley New York 1979, p. 539.
19. H.D. Scobell, K.M. Brobst and E.M. Steele, Cereal Chem. **54,** 1905 (1977).
20. R.T. Yang, L.P. Milligan and G.W. Mathison, "Improved Sugar Separation by High Performance Liquid Chromatography Using Porous Microparticle Carbohydrate Columns," J. Chromatogr. **209,** 316 (1981).
21. D.L. Hendrix, R.E. Lee, J.G. Baust and H. James, "Separation of Carbohydrates Polyols by a Radially Compressed High Performance Liquid Chromatographic Silica Column Modified with Tetraethylene pentamine," J.Chromatogr. **210,** 45 (1981).
22. J. Simmer and J. Puls, J. Chromatogr. **156,** 197 (1978).
23. R.D. Rocklin and C.A. Pohl, "Determination of Carbohydrates by Anion Exchange Chromatography with Pulsed Amperometric Detection," J. of Liquid Chromatogr. **6** (9), 1577 (1983).
24. J.A. Rendleman, "Ionization of Carbohydrates in the presence of Metal Hydroxides and Oxides," Carbohydrates in Solution, Advances in Chemistry, Series Nr. 117, Amer. Chem. Soc. 1973.
25. J.S. Fritz, D.T. Gjerde and R.M. Becker, Anal Chem. **52,** 1519 (1980).
26. O. Samuelson, Z. Anal. Chem. **116,** 328 (1939).
27. C.A. Pohl and J.M. Riviello, Lecture Nr. 108, Rocky Mountain Conference, Denver 1982.
28. R.M. Cassidy and S. Elchuk, Anal. Chem. **54,** 1558 (1982).
29. A.E. Martell and R.M. Smith, "Critical Stability Constants," V3, Plenum Press, New York 1977.

4 Ion Exclusion Chromatography (HPICE)

The introduction of ion exclusion chromatography, which resulted from the work of Wheaton and Bauman [1], can be used to separate weak organic acids and alcohols. As a result of Donnan exclusion, strong, inorganic mineral acids are not retained by the stationary phase and elute within the dead volume time as a single peak. Since they are not affected by Donnan exclusion, undissociated species can penetrate into the pores of the resin. Thus, this separation technique makes use of polar and van-der-Waals interactions between the solutes and the stationary phase.

The major advantage of ion exclusion chromatography lies in the ability to handle samples having very complex compositions. Even injections of dilute soy sauce to separate the contained organic acids do not influence the long term effectiveness of the column.

In combination with ion exchange chromatography (coupled ICE/IC), a separation of organic and inorganic acids can be achieved within some 30 minutes.

Detection is made almost exclusively with conductivity after reducing the background conductivity with the aid of a suitable suppressor system. In terms of sensitivity and selectivity, this method of detection is superior to both RI and shorter wavelength UV detection.

The separation of amino acids is typically performed on resins similar to those used for ICE. Although this separation is predominantly ion exchange, it is considered in this chapter for this reason.

4.1 THE ION EXCLUSION PROCESS

HPICE columns contain totally sulphonated, cation exchange resins with high exchange capacities. The separation process which takes place on the surface of the resin particles involves three different principles:

- Donnan exclusion.
- Steric exclusion.
- Adsorption.

In Figure 4-1, the sulphonic acid groups bound to the surface of the resin particles are shown together with a schematic of the separation processes occurring in an HPICE column. When pure water is pumped through the column, some of the water molecules build hydration spheres around the sulphonic acid groups. The water in these spheres is then in a more ordered state than that predominating in the bulk water.

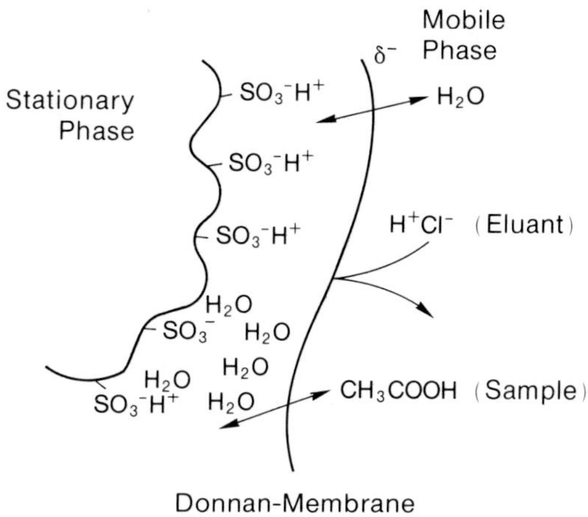

FIGURE 4-1.
Schematic representation of the separation processes on an HPICE column.

In this retention model, the hydration shell is limited by a hypothetical, negatively charged membrane — the Donnan membrane. This membrane is only permeable for neutral compounds. Totally dissociated inorganic acids, such as the dilute hydrochloric acid used as eluant, cannot penetrate this membrane due to the negative charge of the chloride ion. Hence, such dissociated species are completely excluded from the stationary phase and their elution volume is called the exclusion volume, V_e. On the other hand, neutral water molecules can diffuse, unhindered, into and out of the stationary phase pores. The "retention time" of pure water is denoted by the "total permeation volume," V_p, which is observed as a small negative peak when a very sensitive conductivity detector is being used.

When a weak carboxylic acid — such as acetic acid — is injected into the system, the molecules are more or less dissociated, depending on the pH of the eluant. The undissociated form is not affected by Donnan exclusion. Although acetic acid can interact with the stationary phase in a similar manner to water, acetic acid exhibits a greater retention volume than water. This can be explained in terms of adsorption of the acetic acid molecules onto the surface of the resin. Consequently, the separation process for aliphatic carboxylic acids involves both Donnan exclusion and adsorption effects. Furthermore, the longer the aliphatic chain, the greater the retention volume. The addition of acetonitrile to the eluant can reduce the retention times of carboxylic acids by blocking adsorption sites on the resin surface.

Di- and tricarboxylic acids—such as oxalic, citric, etc.—elute between the exclusion and total permeation volumes. In addition to Donnan exclusion, the dominating separation mechanism in this region is steric exclusion. With this process, the retention time of a species is determined by its size. Since the pore size of a resin is fixed by its degree of crosslinking, better resolution in this respect can only be achieved by using another column or a combination of columns.

In general, optimization of the separation of organic acids can be achieved by varying the pH of the eluant since the degree of dissociation (and thus the retention times) is controlled by the pH of the medium.

4.2 STATIONARY PHASES AND ELUANTS

For the analysis of carboxylic and amino acids, four different columns are presently available from Dionex: HPICE-AS1 to AS4. The characteristic properties and applications of these columns are listed in Table 4-1.

TABLE 4-1: Characteristic properties and applications for four ion exclusion columns.

COLUMN	PARTICLE SIZE OF SUPPORT RESIN	DEGREE OF CROSSLINKING	APPLICATION
HPICE-AS1	7.5	9	High performance column for various organic acids.
HPICE-AS2	9.0	8	Column for various organic acids.
HPICE-AS3	15.0	4	Column for organic acids having $pK_a = 2$–4.
HPICE-AS4	6.0	9	Column for amino acids.

All ion exclusion columns contain totally sulphonated cation exchange resins with polystyrene/divinyl benzene as support material. The proportion of divinyl benzene—given in terms of the degree of crosslinking—is of particular importance for the retention behavior of weak organic acids. The extent to which these acids can diffuse into the resin pores is determined directly by the degree of crosslinking. The lower the degree of crosslinking, the longer the retention times of these acids.

The forerunner of the HPICE-AS1 high performance column (backpressure: ca. 58 bar/900 psi at a flow rate of 0.8 cm^3/min) was the HPICE-AS2 column (backpressure: 22 bar/350 psi with a flow rate of 0.8 cm^3/min) that was designed for the first generation of ion chromatographs (Models 10–16). The difference between the two types is simply the particle size of the support material. The HPICE-AS3 column was developed especially for organic acids whose pK_a values lie between 2 and 4. This column is often used in series with an AS1-type column. At the moment, the AS4 column is still used for separations of amino acids. However, a newly developed latex cation exchange column (see Figure 4-2), with a considerably higher efficiency, should soon be in production and will halve the time required for an analysis of complete hydrolysates (see Section 4.8, "New Developments in the Analysis of Amino Acids").

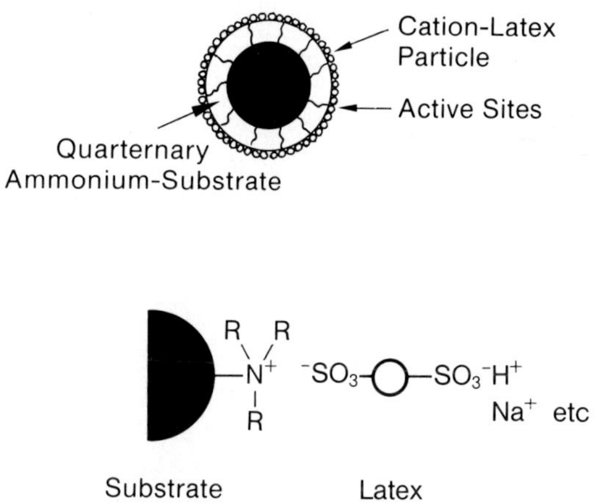

FIGURE 4-2.
Schematic representation of a latex cation exchange resin.

The choice of eluant for ion exclusion chromatography is relatively limited. In combination with a suppressor column in its Ag-form, dilute hydrochloric acid is used for the analysis of a range of carboxylic acids. For the separation of borate and carbonate, pure deionized water can be used as eluant; but since both these acids are only weakly dissociated, their detectivity via conductivity is not particularly sensitive. In order to make very sensitive analyses of borate, a mixed hydrochloride acid/mannitol eluant is preferable. The sugar alcohol serves to complex the boric acid and, thus, gives its solution a higher conductivity.

With the introduction of new types of membrane suppressors for exclusion chromatography, other eluants became more practical. Even though the formerly used hydrochloric acid eluant is compatible with the new membrane suppressors, it leads to a relatively high background conductivity. A good resolution of mono-, di-, and trivalent organic acids can also be achieved using tri-decafluoroheptanoic (perfluorheptanoic acid) at a concentration of ca. 0.001 mol/dm^3. The same concentration of octane sulfonic acid has similar elution properties, although it does lead to a slight tailing of the peaks obtained from aliphatic monocarboxylic acids. In addition, octane sulphonic acid is a suitable eluant for analysis of borate and carbonate. For these analyses, the addition of sugar alcohol can be omitted.

4.3 SUPPRESSORS FOR EXCLUSION CHROMATOGRAPHY

Since the detection of organic acids in ion exclusion chromatography is generally made by measuring their conductivity, it is necessary to employ suppressors to reduce the background conductivity of the eluant. Before the advent of modern membrane suppressors, simple suppressor columns were used that contained a cation exchange resin in Ag-form. With dilute hydrochloric acid, the suppressor reaction can be formulated as follows:

$$\text{Resin} - SO_3^- \, Ag^+ + HCl \rightarrow \text{Resin} - SO_3^- \, H^+ + AgCl \downarrow \qquad (64)$$

Since the organic acid anions are not precipitated by silver, these can proceed, in their original forms, to the conductivity detector. However, this type of suppressor has a number of disadvantages:

- With increasing formation of silver chloride in the suppressor column, an increase in the backpressure and a significant peak broadening are observed. Thus, quantitative evaluation of the chromatogram can be made only via the peak areas in comparison with a reference substance of known concentration.
- In principle, a periodic regeneration of the suppressor column is possible with the aid of a concentrated ammonium hyroxide solution; but, due to the time involved, regeneration is not practicable.
- A continuous regeneration—such as has been used in exchange chromatography for many years—is not possible in conjunction with exclusion chromatography.

These disadvantages led to the development of a cation exchange membrane that can be continually regenerated and used for both ion exclusion as well as ion pair chromatography. This membrane is composed of a sulphonated polyethylene derivative that is resistant to organic solvents. It has a higher permeability for quaternary ammonium bases—such as tetrabutyl ammonium hydroxide—than for hydronium ions. Thus, the mechanism involved here is very different from that described in Chapter 3, Section 3.2, for the ion exchange suppression membranes.

In its regenerated state, the membrane is in its tetrabutyl ammonium form. As the suppressor reaction proceeds, the hydronium ions—associated with the organic acids flowing through the center of the membrane—are exchanged for tetrabutyl ammonium cations. Due to the difference between the equivalent conductivity of the tetrabutyl ammonium and hydronium cations, the resulting salts have a considerably lower conductivity than that of the acids themselves. In the zone of "dynamic equilibrium" a neutralization reaction takes place between H^+ and OH^- to yield the suppressor product — water. The water and the excess tetrabutyl ammonium hydroxide remain in the regenerating medium. The higher background conductivity that results when HCl is used as eluant can be ascribed to the greater equivalent conductivity of the chloride ion as compared to that of the octane sulphonate anion.

TYPICAL OPERATING CONDITIONS

As mentioned above, the suppressor columns that were previously used (cation exchange resins in their Ag-form) were simply thrown away when they became exhausted. Using HCl, with a concentration of ca. 0.001 mol/dm^3, the working life of a cartridge is about 60 hours. In order to not have to considerably raise the pumping pressure to counteract the backpressure buildup after ca. 30 hours use (due to precipitated AgCl), the exhausted half of the column can simply be cut away. Table 4-2 shows the eluants and regenerating media that have been found most suitable for the analysis of organic acids when using the modern AFS-2 membrane suppressors.

TABLE 4-2: Eluants and regenerating media for ion exclusion chromatography.

ELUANT	CONCENTRATION	REGENERATING MEDIA (a)	BACKGROUND CONDUCTIVITY (b)
HCl	$0.5–1.0\times10^{-3}$ mol/dm^3	TBAOH (Tetrabutyl-ammonium-hydroxide) $0.5–1.0\times10^{-3}$ mol/dm^3	50–100 μS/cm
Perfluorheptanoic acid	$0.5–1.0\times10^{-3}$ mol/dm^3		20–40 μS/cm
Octane sulphonic acid	$0.5–1.0\times10^{-3}$ mol/dm^3		25–45 μS/cm

(**a**) The concentration of the regenerating medium should be some ten times more concentrated than the eluant. Flow rate: 2 cm^3/min.
(**b**) The values for the background conductivity given are for a flow rate of 0.8 cm^3/min.

For the analysis of borate and carbonate using an octane sulphonc acid eluant, it is also possible to use an ammonium hydroxide (with a concentration of ca. 0.01 mol/dm^3) regenerating solution.

4.4 THE ANALYSIS OF ORGANIC ACIDS

Using HPICE-AS1 to AS3 columns, it is possible to chromatograph a wide range of organic acids. A selection of organic acids that can be analyzed—listed in order of retention times—is given in Table 4-3.

TABLE 4-3: The elution series of some organic acids that can be determined by ion exclusion chromatography.

↓	↓
Maleic acid	Adipic acid
Oxalic acid	Fumaric acid
Citric acid	Acetic acid
Pyruvic acid	Propionic acid
Malonic acid	Acrylic acid
α-Keto-butyric acid	Isobutyric acid
α-Keto-valeric acid	Butyric acid
Succinic acid	Mandelic acid
Glycolic acid	Pivalic acid
Lactic acid	α-Hydroxy-butyric acid
Formic acid	etc.

Although the strong inorganic mineral acids, under the influence of Donnan exclusion, elute with the dead volume of the system, the retention times of the weaker organic acids can be predicted on the basis of the following criteria:

- Members of a homologous series elute in sequence as a function of their acid strength and solubility in water. For example, the aliphatic monocarboxylic acids: formic acid elutes before acetic acid and the latter before propionic acid, etc. Separations of this nature are very dependent on the distribution of the solute between the mobile and the stationary phases. Van-der-Waals forces between the solutes and the polymeric resin (especially the benzene ring) are particularly responsible for the distribution of the solute between the two phases.
- Generally, dibasic acids elute before monobasic acids. Hence, oxalate elutes before acetate and malonate before propionate, etc.
- The retention time of di- and multibasic acids tends to increase in proportion to the pK_a of the acid.
- Organic acids with branched carbon structures tend to elute before their unbranched isomers. Thus, i-butanoic acid elutes before n-butanoic acid.
- Unsaturated elements in the acid structure lead to longer retention times. Consequently, acrylic acid elutes after propionic acid.
- Due to the strong interactions between their aromatic rings and those of the resin, aromatic acids generally have long retention times.

Figure 4-3 shows a chromatogram obtained by chromatographing some organic acids using dilute HCl (0.001 mol/dm^3) as eluant. The use of a slightly acidic eluant instead of pure distilled water was suggested by Turkelson and Richards [2] in their work concerning the separation of the organic acids formed in the citric acid cycle on an Aminex 50 W − X4 (30–35 μm) cation exchange column. These authors were able to show that the peak form for these acids eluting in pure distilled water tends to be characterized by large peak widths and pronounced tailing effects. The explanation for these phenomena probably lies in the fact that the acids are present both as neutral molecules and as anions. The addition of small amounts of HCl to the eluant, shifts the dissociation equilibria of the acids to the nonionized form and greatly improves the peak forms.

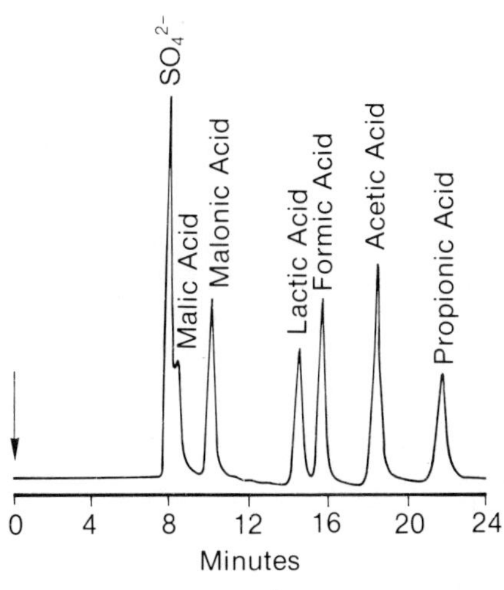

FIGURE 4-3.
Chromatogram of some organic acids. **Column:** HPICE-AS1. **Eluant:** 0.001 mol/dm^3 HCl. **Flow rate:** 0.8 cm^3/min. **Detection:** Conductivity. **Suppressor:** Cation exchange resin in its Ag-form.

The acid concentration in the eluant determines the retention times for di- and multibase organic acids, since the pH determines the degree of dissociation and, thus, the retention behavior. Generally, the resolution improves with increasing acid concentration. This effect is demonstrated in Figure 4-4. The chromatograms were obtained from a mixture of oxalate and sulphate, using different concentrations of HCl in the eluant. The retention of aliphatic monocarboxylic acids can be reduced by adding a small amount (1–3 vol.-%) of acetonitrile to the eluant, since acetonitrile blocks adsorption sites on the surface of the stationary phase.

In comparison to Figure 4-3, the chromatogram shown in Figure 4-5 — obtained using a modern membrane suppressor system — demonstrates the improved resolution of organic acids that can now be achieved. Because of the considerably reduced dead volume of the membrane suppressor, both the peak form and the sensitivity of detection are much improved.

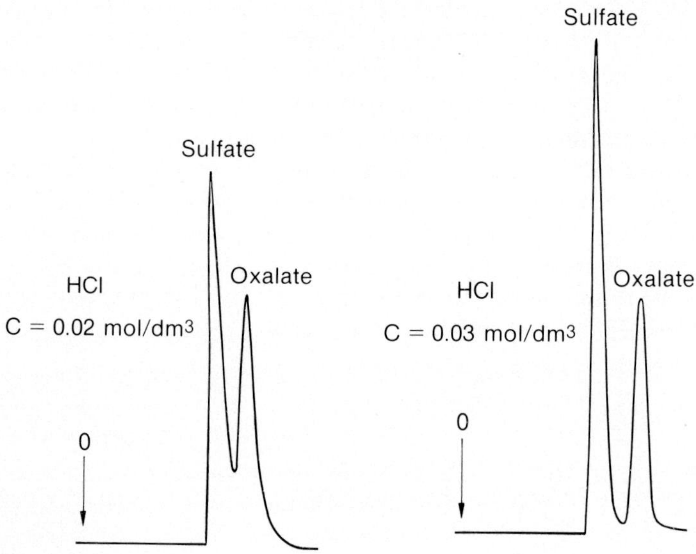

FIGURE 4-4.
Demonstration of the effect of the acid concentration on the retention times with oxalate and sulphate as examples. **Column:** HPICE-AS1. **Flow rate:** 0.8 cm^3/min. **Eluant:** (a) 0.001 mol/dm^3 HCl (b) 0.005 mol/dm^3 HCl. **Detection:** Conductivity. **Suppressor:** Cation exchange resin in its Ag-form.

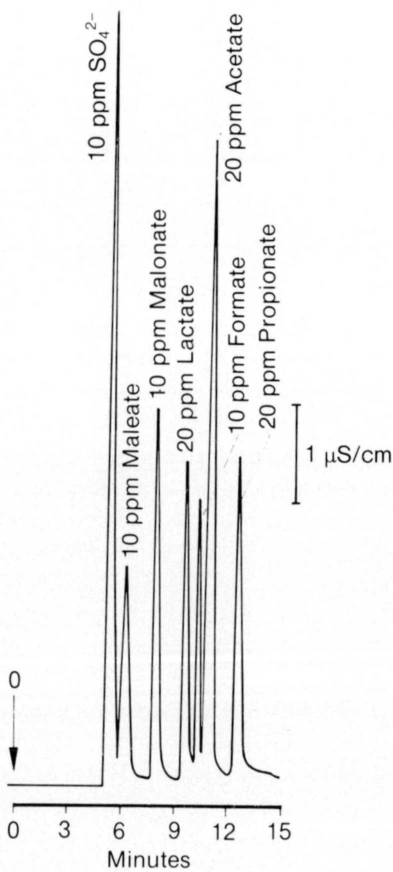

FIGURE 4-5.
Chromatogram of some selected organic acids using a membrane suppressor. **Column:** HPICE-AS1. **Eluant:** 0.001 mol/dm^3 perfluoroheptanoic acid. **Flow rate:** 0.8 cm^3/min. **Detection:** Conductivity. **Suppressor:** AFS-2.

4.5 THE ANALYSIS OF BORATE AND CARBONATE

Boric acid is one of the weakest inorganic acids ($pK_a = 9.23$) and can, depending on the concentration present, be analyzed by one of two different methods. Both techniques are based on a separation using a totally sulphonated cation exchange resin. Because of its low degree of dissociation, boric acid can be detected directly by conductivity only in concentrations between 100 and 4000 ppm. Although pure distilled water can be used as an eluant, it is nevertheless advantageous to use a cation exchange suppressor in its Ag-form to reduce interference from other components in the matrix. The effect of using a membrane suppressor is demonstrated by the chromatograms shown in Figure 4-6 which were obtained with a sample from the primary circulation of a nuclear reactor. It can be seen that the resolution between the substances eluting with the dead volume and borate is considerably improved when a suppressor is employed.

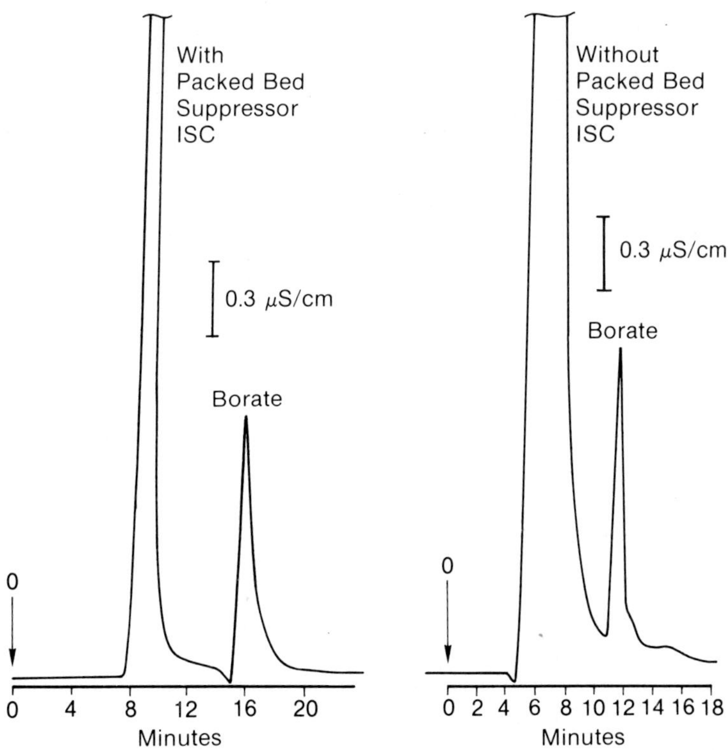

FIGURE 4-6.
Analysis of borate in the presence of a test-matrix with and without a suppressor column. **Column:** HPICE-AS1. **Eluant:** H_2O. **Detection:** Conductivity. **Sample:** 2000 ppm H_3BO_3, 250 ppm Cs^+, 150 ppm NO_3^-, 40 ppm I^-, 20 ppm K^+, 10 ppm Cl^-, 10 ppm Ba^+, 5 ppm NH_4^+, Ce^{4+} and La^{3+} and 2 ppm Li^+.

Up to now, the analysis of very small amounts of borate has been achieved using a mixture of dilute HCl and mannitol as eluant and a cation exchange column in its Ag-form as the suppressor column. With the introduction of the AFS-2 membrane suppressors, the disadvantages of the more traditional suppressor could be circumvented and the sensitivity of the analyses considerably improved. Figure 4-7 shows a chromatogram obtained from a sample containing 1 ppm borate. The limit of detection for borate is approximately 100 ppb.

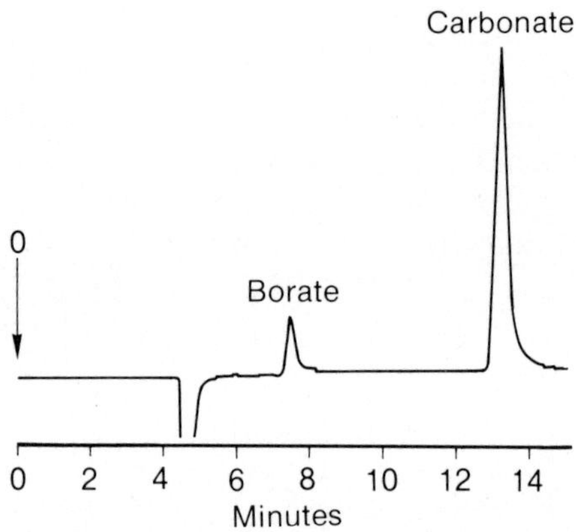

FIGURE 4-7.
Analysis of borate and carbonate using a membrane suppressor. **Column:** HPICE-AS1. **Eluant:** 0.001 mol/dm^3 octane sulphonic acid. **Regenerating medium:** 0.01 mol/dm^3 NH_4OH. **Detection:** Conductivity. **Suppressor:** AFS-2. **Sample:** 10 ppm H_3BO_3, 50 ppm CO_3^{2-}.

As can be seen from Figure 4-7, it is possible to analyze carbonate and borate simultaneously when using octane sulphonic acid as eluant and ammonium hydroxide as the regenerating medium. The detection limit for carbonate is approximately 1 ppm.

4.6 COUPLED HPICE/HPIC

The concept of combining ion exclusion and ion exchange chromatography was introduced by Rich and his coworkers [3] and serves to improve the selectivity of separations of organic and inorganic acids in complex matrices. Schematically, Figure 4-8 shows how the method works in principle.

After injection, the sample being analyzed flows onto an ion exclusion column that separates the organic acids with higher pK_a values from the inorganic acids. The latter elute from this column is a single peak. The effluent from the HPICE column then travels through the suppressor column, where the background conductivity of the eluant is reduced and interfering cations are exchanged for protons. The selectivity of the separation system can be increased by including—via column switching—a concentrator column packed with a low capacity, anion

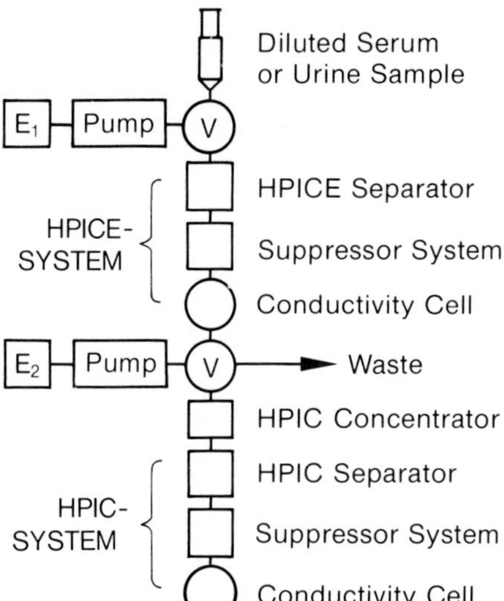

FIGURE 4-8.
Scheme showing the way a coupled ICE/IC system works.

exchange resin. On the concentrator column, one or more of the peaks (separated by HPICE) is collected and then further separated on the anion exchange column (connected in series), using carbonate or hydrogen carbonate as eluant. In effect, the HPICE column is, thus, an online extraction unit. By choosing an eluant of suitable ionic strength, a wide range of inorganic and organic anions can be analyzed with a minimum effort.

The coupled ICE/IC technique has been used successfully by Rich and his coworkers [4] for separating pyruvate and lactate in blood serum. Simple ion exchange chromatography has limited usefulness for the separation of organic acids in biological samples, since the retention behavior of organic acids is very similar to some inorganic anions. Using coupled ICE/IC, this problem ceases to hinder the analysis. Organic acids, which have similar retention times with respect to the HPICE phase, can be separated easily in the anion exchange column. Thus, α-ketovaleric acid, α-ketoglutaric acid, and pyruvic acid cannot be separated by HPICE under usual chromatographic conditions. In contrast, the capacity factors of these compounds in an anion exchange column are very different. Figure 4-9 shows the chromatogram of a normal human serum.

In Figure 4-9, the peaks identified by letters are due to several organic acids which have similar pK_a values and hydrophobicity. If the peak labeled "A" is first concentrated and then analyzed using anion exchange chromatography, the chromatogram shown in Figure 4-10 is obtained. As can be seen, this single peak is caused by both nitrate and sulphate ions.

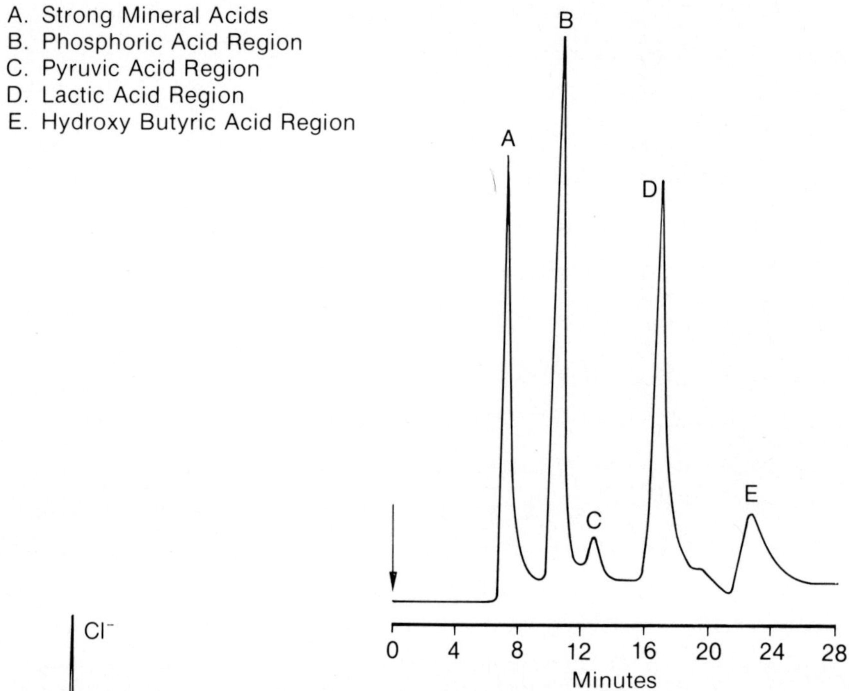

FIGURE 4-9.
HPICE chromatogram of a human serum. **Column:** HPICE-AS2. **Eluant:** 0.01 mol/dm^3 HCl. **Detection:** Conductivity. **Suppressor:** Cation exchange resin in its Ag-form.

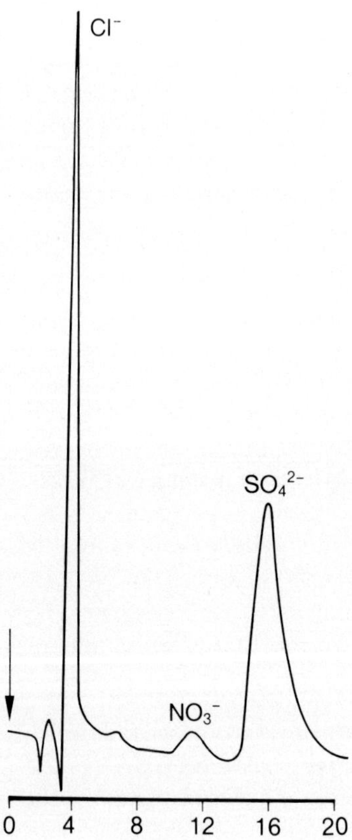

FIGURE 4-10.
HPIC chromatogram of the peak labelled "A" in Figure 4-9. **Column:** HPIC-AS1. **Eluant:** 0.003 mol/dm^3 $NaHCO_3$, 0.0024 mol/dm^3 Na_2CO_3. **Detection:** Conductivity. **Suppressor:** ASC-2.

The limit of detection for organic acids is determined by the concentration ratios of the individual components in the sample; but, using the method described above, it is in the low ppb range.

4.7 THE ANALYSIS OF ALCOHOLS

Mono- and multi-alcohols with short carbon chains can be separated on totally sulphonated cation exchange resins using perchloric acid as eluant. Figure 4-11 shows a typical chromatogram from such a separation.

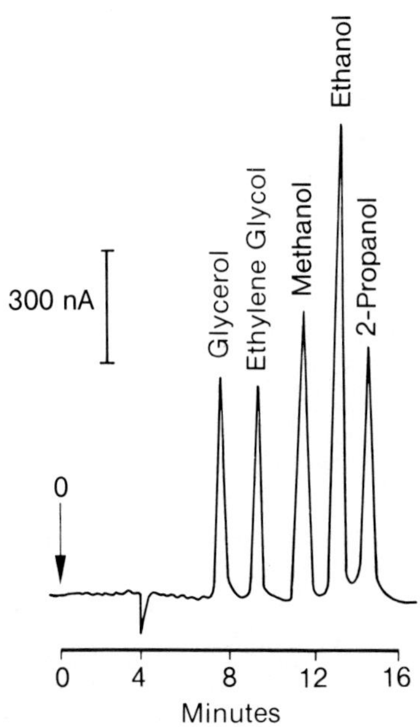

FIGURE 4-11.
Analysis of single and multifunctional alcohols. **Column:** HPICE-AS1. **Eluant:** 0.1 mol/dm^3 $HClO_4$. **Detection:** Pulsed amperometry at a Pt working electrode.

The key substance in this mixture is glycerin. Due to its high boiling point, glycerin cannot be separated by GC. Although the mono-alcohols can be analyzed by HPLC techniques, the advantage of ion chromatography lies in the ability to simultaneously analyze multi-alcohols. Furthermore, detection by pulsed amperometry is much more sensitive than the RI detectors usually used for HPLC.

4.8 THE ANALYSIS OF AMINO ACIDS

Ion chromatography with totally sulphonated cation exchange resins is the most widely used technique for the separation of amino acid mixtures. One method, first reported by Spackman, Stein, and Moore [5], uses two exchange columns. The first separates the more acid and neutral amino acids. The second separates the more basic amino acids. Due to uncontrollable factors, unstable baselines and several artifacts made it impossible to use a single exchange phase for the separation of amino acids. This type of analysis could only be used after new types of buffer systems had been developed. Today, two types of separation techniques are available:

- Gradient elution according to the method of Piez and Morris [6].
- Sequential elution according to the method of Hamilton [7].

Gradient elution involves steadily increasing the concentration of cations in the eluant while holding the pH value constant; whereas, the reverse procedure is called *sequential elution.* The latter method is preferred, since the analysis times are shorter and problems with the stability of the baseline are more easily solved.

ELUANTS FOR AMINO ACID ANALYSIS

Generally, the separation of amino acids involves using several buffer solutions with increasing cation concentration and increasing pH. Dus and his coworkers [8] employed a sequence of four citrate buffers whose pH varied between 3.2 and 10.9 and whose Na^+ concentration varied between 0.2 and 1.4 mol/dm^3. As an alternative, Hare [9] developed an elution technique based on a constant cation concentration (0.2 mol/dm^3 Na^+) and in which the pH of the system was varied between 3.2 and 10.1. The Pico-buffer system, developed by Benson [10], uses a series of buffer solutions having similar pH values but varying cation concentrations. In contrast to systems involving pH gradients, this procedure has the advantage of leading to more stable baselines. Using this procedure, the elution of amino acids is determined solely by the concentration of Na^+ or Li^+ ions in the eluant.

The HiPhi buffer systems, recommended by Dionex and similar to the systems developed by Dus, et al., are listed in Table 4-4.

Whereas the sodium systems are more suitable for the separation of complete hydrolysates, the lithium systems are more suitable for the rapid analysis of amino acids in biological liquids.

TABLE 4-4: The composition of the buffer solutions recommended by Dionex for the separation of amino acids.

TYPE OF ELUANT	VALUE	CONCENTRATION OF CATIONS
Na-A	3.25 pH	0.2 mol/dm^3
Na-B	4.25 pH	0.2 mol/dm^3
Na-C	7.40 pH	1.0 mol/dm^3
Li-A	2.75 pH	0.238 mol/dm^3
Li-B	2.65 pH	0.238 mol/dm^3
Li-C	3.60 pH	0.340 mol/dm^3
Li-D	4.17 pH	0.667 mol/dm^3
Li-E	5.30 pH	0.643 mol/dm^3

The chemicals used for preparing the individual mixtures should be of the highest purity. In particular, ammonium and amine impurities — present in laboratory air and very soluble in the buffer solutions — strongly affect subsequent analyses. Ammonia in the ammonium ion form elutes very slowly and, if the sample to be analyzed contains noticeable amounts of ammonium, this will elute as a sharp peak concurrent with the strong basic amino acids. If, on the other hand, the eluant is contaminated with ammonia, the ammonium ions are continually bound onto the stationary phase. Since the ammonium ion has a strong affinity for the stationary phase, these ions travel very slowly through the column and elute to give a broad and diffuse signal which is observed as baseline instability. It is for these reasons that buffer solutions, intended to be used for the separation of amino acids by ion chromatography, should be stored under a blanket of inert gas. Suitable solvent reservoirs are commercially available.

To prevent the frequent occurring growth of microorganisms in buffer solutions, the Dionex HiPhi buffer solutions contain 1.0 g/dm^3 phenol.

THE FACTORS DETERMINING RETENTION TIMES

Amino acids are amphoteric; i.e., they may be present in a given solution in either their anionic or cationic forms. In buffer solutions having a low pH, the equilibrium

$$H_2N - CH_2 - COOH + H^+ \rightleftharpoons H_3N^+ - CH_2 - COOH \tag{65}$$

is shifted to the right (as written). In their cationic form, amino acids can interact with the sulphonate groups of the cation exchange resin. As the pH of the eluant decreases, the equilibrium shown in Equation 65 is shifted further to the right and the concentration of exchangeable protons increases. Thus, when the eluant has a low pH, amino acids are more strongly retained by the stationary phase. On the other hand, as the pH of the eluant increases, the equilibrium in Equation

65 shifts to the left and the neutral amino acid molecules produced do not interact with the stationary phase and, hence, elute very rapidly.

Even though the pH of the buffer solution used as eluant is an important factor for the retention behavior, the concentration of cations in the eluant also plays a role in determining the retention times of amino acids. Since the buffer solutions used are generally based on sodium citrate, sodium—as well as amino acid cations—can interact with the exchange functions on the stationary phase. The competing equilibria can be formulated as Equation 66:

$$Na^+ + Resin^n\ SO_3^-\ A^+ \rightleftharpoons Resin^n\ SO_3^+\ Na^+ + A^+ \qquad (66)$$

A^+: Amino acid in its cationic form.

Although the amino acids have a much stronger affinity for the stationary phase than the sodium cations, the latter are present in considerably greater concentration and are continually replenished. Therefore, the equilibrium shown by Equation 66 is shifted to the right and the amino acids are released more rapidly from the stationary phase when the sodium ion concentration is greater. Figure 4-12 shows a typical chromatogram of the more important amino acids contained in a cattle food matrix.

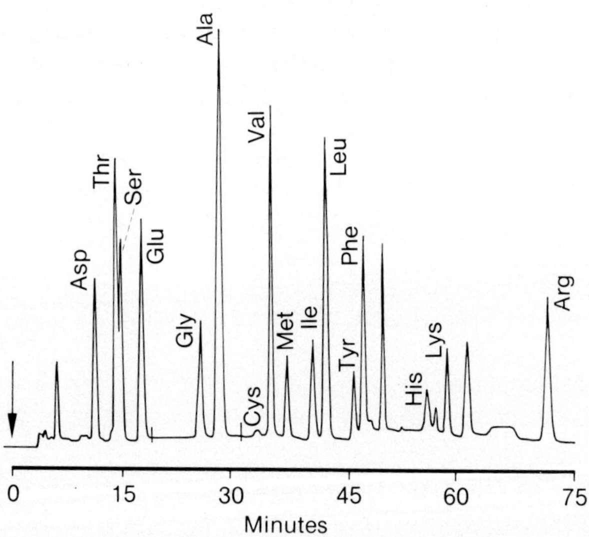

FIGURE 4-12.
A separation of the amino acids in cattle food. **Column:** HPICE-AS4. **Eluant:** HiPhi buffer A, B and C. **Temperature:** 40/65°C. **Flow rate:** 0.2 cm^3/min. **Detection:** Fluorescence after reaction with o-phthalaldehyde.

In addition to the pH and the cation concentration of the eluant, the column temperature and the flow rate play a small role in determining the retention behavior of amino acid solutes. Normally, the column temperature is adjusted to 40°C at the start of an analysis. After valine has eluted, it is raised to 65°C to decrease adsorption effects. Due to the low permeability of the resins used, only low flow rates (0.2 cm^3/min) can be used without harming the column packing.

NEW DEVELOPMENTS IN THE ANALYSIS OF AMINO ACIDS

The conventional techniques for analyzing amino acids, described in the previous section, have a number of disadvantages which can be summarized as follows:

- The cation exchange resins used up to now are totally sulphonated materials. Therefore, the kinetic processes are relatively slow, since the sample molecules can diffuse into the bulk of the stationary phase.
- Totally sulphonated materials are mechanically weak and have low permeabilities. Thus, with a high column backpressure, only very low flow rates can be employed. Should greater flow rates occur accidentally, the quality of the column packing and the separation performance will deteriorate. Consequently, shorter analysis times can only be achieved by an appropriate change in eluant.
- Packing high quality columns is both time consuming and difficult.
- Analyses employing a citrate buffer system with different counterion cannot be carried out on a single column, since this would lead to voids developing in the column bed.
- Analyses times of more than 80 minutes are not acceptable for routine analyses.

For the above reasons, the development of a cation exchange resin with better separation properties is the aim of intensive research at the present time. There are two important goals:

- The column should be capable of separating a complete hydrolysate within 30 minutes.
- The column backpressure should not exceed 65 bar (1000 psi) with a flow rate between 1 and 2 cm^3/min.

The development work has followed two different approaches: separation of amino acids using either *anion* or *cation* exchange resins. The basic principle of both types of column is a latex construction. This has already been realized for the anion exchange material. Column packings of this type have a mechanically stable "core" whose surface is functionalized. The latex particles having the actual exchange functions are bound to this core electrostatically. Based on work performed in Dionex research laboratories, it is now possible to make a similar material with cation exchange properties. The surface of the "core" of the latter material is aminated and has totally sulphonated latex particles bound to it. The pellicularity; i.e., the surface functionality of a resin, is the prerequisite for the successful accomplishment of one of the two aims described.

Columns which exhibit low backpressures can be made using pellicular resins which have a high permeability. Another advantage of the latex cation exchange particles — in direct analogy to the latex anion exchange materials — is the short diffusion paths which result and the consequent high efficiency of such columns. Initial experiments with a prototype system suggest that the quality of the column packing is not influenced, even when the counterion of the citric buffer is changed from Na^+ to Li^+. In contrast to the totally sulphonated cation exchange resins, the newly developed latex materials have considerably lower capacities. In particular, this parameter is influenced by the size of the latex particles which is between 0.1 and 1.5 μm. The larger the latex particles, the greater the capacity of the column. Latex cation exchange columns are characterized by much higher selectivities, due to the hydrophobicity and functionality of the latex particles.

Figure 4-13 shows a chromatogram from the separation of the amino acids which elute relatively rapidly using a latex cation exchange column. The high selectivity of the column is particularly notable. Typical of this improved technique is that it can be carried out isocratically at room temperature.

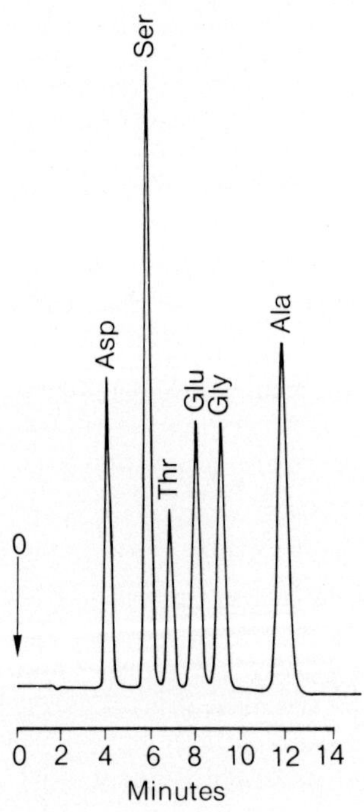

FIGURE 4-13.
A separation of some amino acids on a newly developed latex cation exchange column. **Column:** Prototype from Dionex. **Eluant:** Aqueous HCl; pH 3.2. **Detection:** Fluorescence after reaction with o-phthalaldehyde. **Sample:** 50 μl of a solution containing 125 pmol each of asparagine, serine, threonine, glutamine acid, glycine and alanine.

The six amino acids responsible for the peaks shown in Figure 4-13 are difficult to separate with conventional techniques so that the performance of the new columns can be effectively judged. The simple composition of the eluant is especially worth mentioning. With dilute hydrochloric acid (pH 3.2), it is possible to completely separate all signals within 14 minutes. In comparison, using conventional exchange columns (see Figure 4-12), the same analysis requires some 30 minutes.

The parameter that has the most influence on the retention times, is the proton concentration in the mobile phase. Most affected by this parameter is glutamic acid, due to its low isoelectric point. Up to now, no techniques have been developed for separating amino acids using anion exchange columns. Almost certainly, the reason for this is the difficulty associated with the synthesis of directly aminated materials also having suitable chromatographic properties. However. the latex anion exchange materials could provide the basis of an alternative technique. From the chromatogram shown in Figure 4-14, it can be seen that the elution sequence is essentially reversed when compared to separations on cation exchange resins. Arginine and lysine elute most rapidly. Threonine and serine are considerably better separated, but alanine and threonine coelute. Under the conditions used to obtain the chromatogram shown in Figure 4-14, the lack of resolution of the signals for isoleucine and leucine can be avoided simply by slightly changing the pH of the eluant (see Figure 4-15). Another prototype hyper-crosslinked, latex anion exchange resin has a somewhat different selectivity. With this column, as can be seen from Figure 4-16, the problem of separation of threonine and alanine can be solved; but, with this column, valine and glycine coelute.

From the examples discussed above, it becomes evident that the development of suitable anion and cation exchange materials is a long way from being complete. The already developed cation exchange materials do have the desired selectivities, so that future development work will concentrate on improving the capacity of these columns.

In contrast, the anion exchange materials seem to have a sufficient capacity, but their selectivity requires further improvement. Nevertheless, the examples given above do suggest that commercially available columns of suitable quality can be expected in the not too distant future.

Since the writing of the German version of this handbook, both anion (AS-8) and cation (AS-4) columns have become commercially available for the separation of amino acids.

FIGURE 4-14.

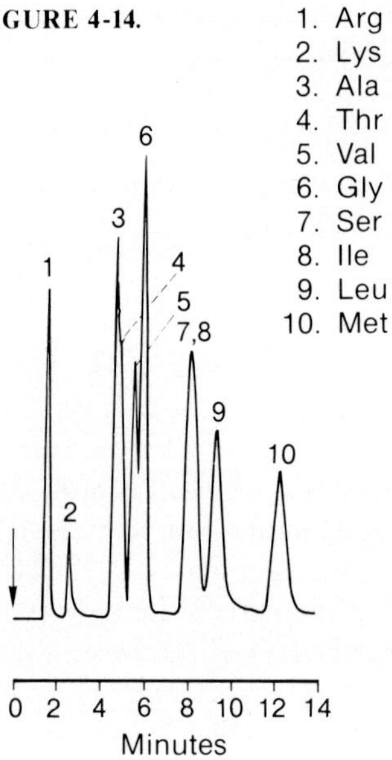

FIGURE 4-15.

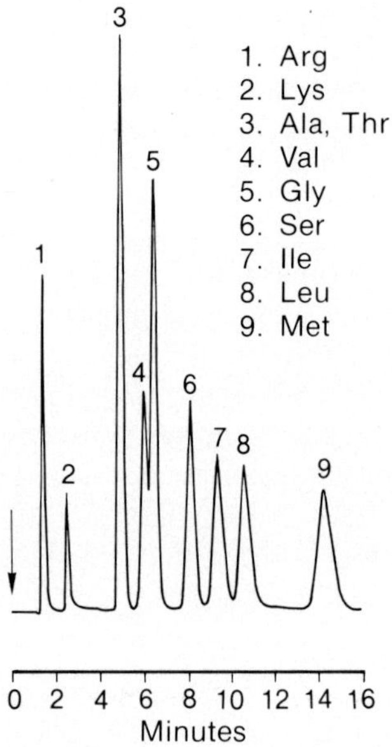

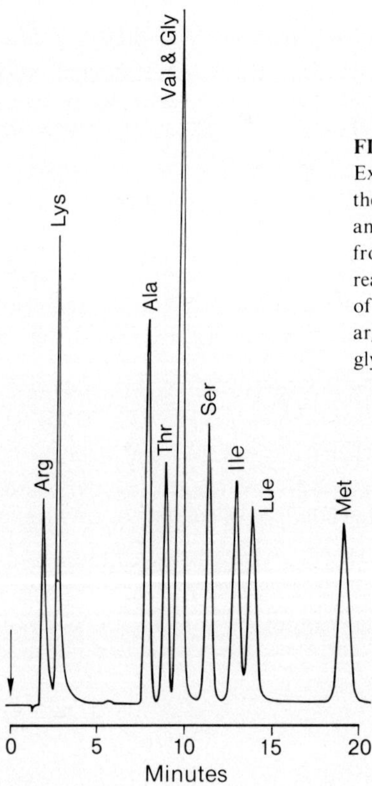

FIGURE 4-16.

FIGURES 4-14, 4-15 AND 4-16.
Examples of the separation of amino acids using the newly developed hyper-crosslinked latex anion exchange columns. **Column:** Prototypes from Dionex. **Detection:** Fluorescence after reaction with o-phthalaldehyde. **Sample:** 50 μl of a solution containing 125 pmol of each of arginine, lysine, alanine, threonine, valine, glycine, serine, isoleucine, leucine and methionine.

LITERATURE

1. R.M. Wheaton and W.C. Bauman, Ind. Eng. Chem. **45,** 228 (1953).
2. V.T. Turkelson and M. Richards, Anal. Chem. **50,** 1420 (1978).
3. W.E. Rich, F. Smith, L. McNeill, and T. Sidebottom, "Ion Exclusion Coupled to Ion Chromatography: Instrumentation and Application," in Ion Chromatographic Analysis of Environmental Pollutants, Vol. 2, Ann Arbor Science, Ann Arbor, Michigan (1979), p. 17.
4. W.E. Rich, E.L. Johnson, L. Lois, B.E. Stafford, P.M. Kabra, and L.J. Marton, "Organic Acids by Ion Chromatography," Chapter 17, Liquid Chromatography in Clinical Analysis, editor: L.J. Marton, P.M. Kabra, The Human Press, Inc. 1981.
5. D.H. Spackman, W.H. Stein, and S. Moore, Anal. Chem. **30,** 1190 (1958).
6. K. Piez and L. Morris, Anal. Biochem. **1,** 187 (1960).
7. P. Hamilton, Anal. Chem. **35,** 2055 (1963).
8. K. Dus, S. Lindroth, R. Pabst, and R. Smith, Anal. Biochem. **14,** 41 (1966).
9. P.E. Hare, Space Life Sciences **3,** 354 (1972).
10. J.R. Benson, U.S. Patent nr. 3,686,118 (1972).
11. A. Woodruff, C. Pohl, S. Papanu, C. Lincoln, and D. Fan, "Recent Developments in Amino Acid Analysis," Paper Nr. 75, 35th Pittsburg Conference, Atlantic City, March 1984.

5 Ion Pair Chromatography (MPIC)

Ion pair chromatography has become increasingly important as an alternative to ion exchange chromatography since it can separate and quantify both cations and anions.

Haney and his coworkers [1,2], Waters Associates [3], as well as Knox and his coworkers [4,5], have found that by adding lipophilic ions (such as alkyl sulphonic acids or quaternary aliphatic amines) to the mobile phase, solute ions with opposite charges can be separated on the common reversed phase silica columns. The term "**r**eversed **p**hase **i**on **p**air **c**hromatography" (RPIPC) has become established for identifying this technique. The term "**m**obile **p**hase **i**on **c**hromatography" (MPIC) is a patented technique essentially the same as RPIPC, but includes detection via conductivity after reduction of the background conductivity using a suppressor system. The stationary phase used for this technique is a neutral macroporous polystyrene-divinyl benzene resin with a nonpolar character.

The physicochemical phenomena on which the retention mechanism is based have not yet been fully explained. The lack of a well defined mechanistic explanation of the processes involved are reflected in the variety of terms that have been applied to such separations. At present, two different hypotheses — both lacking indubitable experimental corroboration — are being discussed. Horvath and his coworkers [6, 7] propose that the solute ions form neutral ion pairs with the lipophilic ions in the aqueous phase and these are adsorbed onto the neutral stationary phase. On the other hand, Huber, Hoffmann, and Kissinger [8, 9, 10] support the alternative, ion exchange model, whereby the lipophilic ions are adsorbed onto the surface of the stationary phase and turn it, effectively, into an ion exchange material. However, both hypotheses represent extreme cases and, as such, each fails to explain the retention processes under all chromatographic conditions. Furthermore, the development of a single retention model that takes all experimental results into account is not possible, simply on the basis of chromatographic data.

Obtaining a better understanding of the very complex mechanisms involved is hampered by the lack of physicochemical data in the literature and the difficulties involved in obtaining the relevant equilibrium constants.

5.1 SUMMARY OF THE EXISTING RETENTION MODELS

According to the *ion pair formation* model [6,11], the solute ion E interacts with the lipophilic ion H and forms a complex, EH. This complex can be bound reversibly on the nonpolar, stationary phase L in the form LEH. The equilibria involved can be formulated as:

$$E + H \xrightleftharpoons{K_1} EH \tag{67}$$

$$EH + L \xrightleftharpoons{K_2} LEH \tag{68}$$

The corresponding equilibrium constants can be expressed as given in Equations 69 and 70, indicating the concentrations of the individual species in the mobile and stationary phases by the subscripts m and s, respectively:

$$K_1 = [EH]_m / [E]_m \cdot [H]_m \tag{69}$$

$$K_2 = [LEH]_s / [EH]_m \cdot [L]_s \tag{70}$$

where $[L]_s$ is a measure of the available, free surface of the stationary phase.

In contrast, according to the *ion exchange model* [8, 9, 10], the lipophilic ions are bound to the surface of the stationary phase to form LH. Thus, the nonpolar resin is converted into a dynamic ion exchange material. The solute ion E can then interact with LH as follows:

$$H + L \xrightleftharpoons{K_3} LH \tag{71}$$

$$LH + E \xrightleftharpoons{K_4} LEH \tag{72}$$

for which the equilibrium constants can be formulated as:

$$K_3 = [LH]_s / [H]_m \cdot [L]_s \tag{73}$$

$$K_4 = [LEH]_s / [LH]_s \cdot [E]_m \tag{74}$$

Assuming that the solute and lipophilic ions interact with the stationary phase independent of one another, a very complex system of equilibria (described by Horvath [12] and shown here in Figure 5-1) can be established.

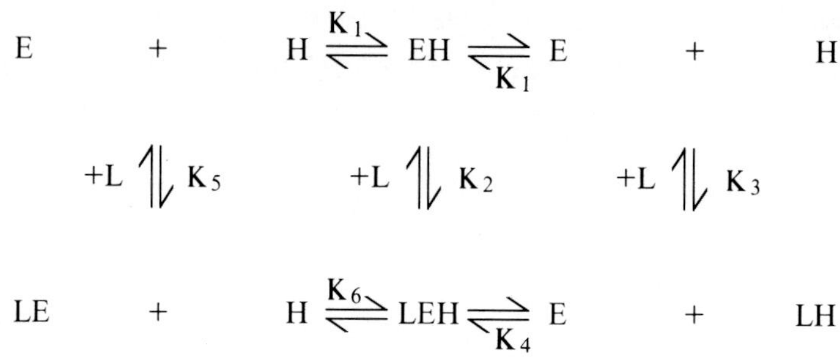

FIGURE 5-1.
Schematic representation of the equilibria operating during ion pair chromatography on a nonpolar stationary phase and with lipophilic ions in the mobile phase.

In this case, Equations 75 and 76 need to be added to the sequence of equations given above:

$$[E] + [L] \overset{K_5}{\rightleftharpoons} [LE] \tag{75}$$

$$[LE] + [H] \overset{K_6}{\rightleftharpoons} [LEH] \tag{76}$$

and

$$K_5 = [LE]_s / [E]_m \cdot [L]_s \tag{77}$$

$$K_6 = [LEH]_s / [LE]_s \cdot [H]_m \tag{78}$$

Using the definition of the capacity factor k′:

$$k' = \Phi \cdot K \tag{79}$$

Where Φ is the phase volume ratio and K is the distribution coefficient, it is possible to derive the capacity factor for the solute ion E. Thus,

$$k' = \Phi \cdot \frac{[LEH]_s \cdot [LE]_s}{[E]_m \cdot [EH]_m} \tag{80}$$

If the stationary and mobile phases are at equilibrium, then the concentration of lipophilic ions is constant. With

$$[E] << [H]$$

only a fraction of the lipophilic ions are complexed with the stationary phase and the concentration of these ions is essentially constant:

$$[H]_m \cong [H] \tag{81}$$

If one further assumes that the proportion of the solute ions bound to the stationary phase is very small and the total surface area of the stationary phase L is fixed, then

$$[L]_T = [L]_s + [LH]_s \cong [L] \tag{82}$$

is valid. In the case that the lipophilic and solute ions form ion pairs in the mobile phase and interact as the ion pair with the stationary phase, then the capacity factor can be formulated as a function of H and L by using the relationship given in Equations 69, 70, 73, 74, 81, and 82. To achieve this, Equations 69, 70, 73, and 74 are rearranged as follows:

$$[LE]_s = K_1 \cdot [E]_m \cdot [L]_s \tag{83}$$

$$[E]_m = [EH]_m / K_2 \cdot [H]_m \tag{84}$$

$$[L]_s = [LH]_s / K_3 \cdot [H]_m \tag{85}$$

$$[LEH]_s = K_4 \cdot [EH]_m \cdot [L]_s \tag{86}$$

By incorporating Equations 83–86 in Equation 80 and separating the term $[L]_s/[E]_m$ one obtains:

$$k' = \Phi \cdot \frac{[L]_s}{[E]_m} \cdot \frac{K_4\,[EH]_m + K_1 \cdot [E]_m}{1 + K_2 \cdot [H]_m} \tag{87}$$

This expression can be expanded by multiplying the top and bottom by the term $(1 + K_3 \cdot [H]_m)$ to give:

$$k' = \Phi \cdot \frac{[L]_s}{[E]_m} \cdot \frac{(1 + K_3 \cdot [H]_m) \cdot (K_4 \cdot [EH]_m + K_1 \cdot [E]_m)}{(1 + K_2 \cdot [H]_m) \cdot (1 + K_3 \cdot [H]_m)} \tag{88}$$

which, with Equations 84 and 85, gives:

$$k' = \Phi \cdot \frac{K_2}{K_3}\,\frac{[LH]_s}{[EH]_m} \cdot \frac{(1 + K_3 \cdot [H]_m) \cdot (K_4 \cdot [EH]_m + K_1 \cdot [E]_m)}{(1 + K_2 \cdot [H]_m) \cdot (1 + K_3 \cdot [H]_m)} \tag{89}$$

and

$$k' = \Phi \cdot [LH]_s \cdot \frac{\left(\dfrac{K_2}{K_3} + K_2 \cdot [H]_m\right) \cdot \left(K_4 + \dfrac{K_1 \cdot [E]_m}{[EH]_m}\right)}{(1 + K_2 \cdot [H]_m) \cdot (1 + K_3 \cdot [H]_m)} \tag{90}$$

By incorporating Equation 84 and rearranging, one can then obtain:

$$k' = \Phi \cdot [LH]_s \cdot \frac{\dfrac{K_2 \cdot K_4}{K_3} + \dfrac{K_1}{K_3 \cdot [H]_m} + K_2\,K_4 \cdot [H]_m + K_1}{(1 + K_2 \cdot [H]_m) \cdot (1 + K_3 \cdot [H]_m)} \tag{91}$$

and

$$k' = \Phi \cdot \frac{[LH]_s \cdot \left(\frac{K_2 K_4}{K_3} + \frac{K_1}{K_3 \cdot [H]_m}\right) + [LH]_s \cdot (K_2 K_4 \cdot [H]_m + K_1)}{(1 + K_2 \cdot [H]_m) \cdot (1 + K_3 \cdot [H]_m)} \qquad (92)$$

Incorporation of Equation 85 then leads to:

$$k' = \Phi \cdot \frac{K_2 K_4 \cdot [H]_m \cdot [L]_s + K_1 \cdot [L]_s + [LH]_s \cdot (K_2 K_4 \cdot [H]_m + K_1)}{(1 + K_2 \cdot [H]_m) \cdot (1 + K_3 \cdot [H]_m)} \qquad (93)$$

By separating the $[L]_s$ terms and taking Equation 82 into account, the desired relationship for k' can now be obtained:

$$k' = \Phi \cdot [L] \cdot \frac{K_1 + K_2 K_4 \cdot [H]}{(1 + K_2 \cdot [H]) \cdot (1 + K_3 \cdot [H])} \qquad (94)$$

(Ion Pair Model)

According to the dynamic exchange model and using the relationships given by Equations 69, 70, 73, 77, and 80–82, an alternative expression for the capacity factor k' (Equation 95), can be obtained in a similar way to that described in detail above.

$$k' = \Phi \cdot [L] \cdot \frac{K_1 + K_3 K_5 \cdot [H]}{(1 + K_2 \cdot [H]) \cdot (1 + K_3 \cdot [H])} \qquad (95)$$

(Dynamic Ion Exchange Model)

Both Equation 94 and Equation 95 demonstrate the dependence of the capacity factor k' on the concentration of lipophilic ions. The general form of this dependence is given by:

$$k' = (k_0 + B[H]) / (1 + K_2[H]) \cdot (1 + K_3[H]) \qquad (96)$$

k_0: The capacity factor for the solute ions when no lipophilic ions were involved in the retention process.

K_2: The equilibrium constant for the appropriate ion pair formation.

K_3: The formation of the complex between the stationary phase and the lipophilic ions.

From Equations 94 and 95, it can be seen that the constant B is the product of two rate constants.

A plot of k' as a function of $[H]$ gives a parabola according to Equation 96 with $1/K_3 \cdot [H] << 1/K_2$. Knox and his coworkers have obtained similar relationships experimentally [5,12].

In 1979, Bidlingmeyer and his coworkers [13,14] introduced a third model — the "ion interaction" model — into the discussion. This model was derived from conductivity measurements that excluded the possibility of ion pair formation in the mobile phase. This model for the retention processes occurring during RPIPC (also used by Pohl [15] to interprete the retention mechanism on an MPIC-phase), neither assumes the formation of ion pairs nor is it based on classical ion exchange processes.

Due to the nonpolar character of the stationary phase and the polar character of the mobile phase, a high surface tension exists at the boundary between the phases. This leads to the stationary phase having an affinity for components in the mobile phase which are capable of reducing the surface tension. Such components include organic solvents, surfactants and their counterions, as well as quaternary ammonium bases. According to the "ion interaction" model, an electrical double layer is built up on the surface of the stationary phase. Figure 5-2 schematically shows this concept with an analysis of surface-inactive anions as an example.

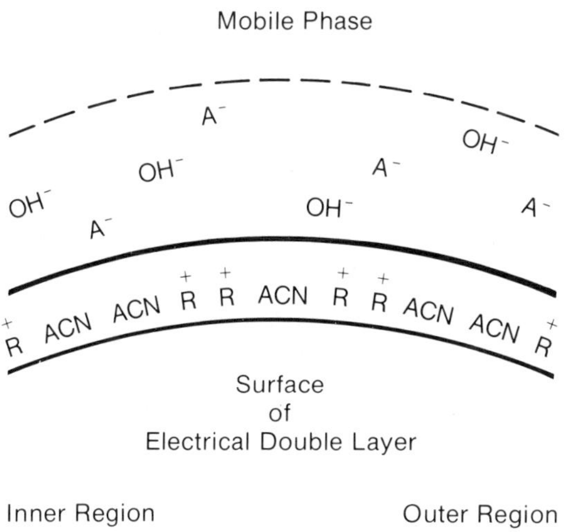

FIGURE 5-2. Scheme showing the electrical double layer formed during the separation of surface active anions.

In the inner region, the lipophilic ions that have been added to the mobile phase (e.g., tetrabutyl ammonium cations) and the organic modifier (ace tonitrile) are adsorbed onto the surface of the nonpolar, stationary phase. Since the lipophilic cations have the same charge, only a small portion of the surface of the stationary phase can be occupied by these ions due to repulsion forces. The corresponding counterions (when conductivity detection is being used, these are normally OH^-) and the anions (A^-) being analyzed, make up the diffuse, outer region. If the concentration of lipophilic ions in the mobile phase is increased, the concentration of these ions adsorbed on the surface of the stationary phase will also increase due to the dynamic nature of the equilibrium involved. The selectivity of the system is derived from the fact that the charge density in the outer region of the double layer is dependent on the degree of hydration of the counterions. Counterions with low hydration enthalpies—such as iodide—lead to a greater charge density than anions—such as OH^-—that have greater enthalpies associated with their hydration. As a result, the charge density of the lipophilic ions on the resin surface can be increased. This, in turn, leads to a further decrease in the surface tension. The consequence of these processes can be summarized by: the ions having the smallest enthalpy of hydration will have the longest retention times.

By analogy, the separation of surfactant cations can also be mechanistically interpretated. In this case, lipophilic anions are adsorbed onto the resin surface and the cations being analyzed are retained in the outer region of the electrical double layer.

In contrast, surface active ions can penetrate the inner region of the double layer and become adsorbed onto the resin surface. For this class of substance, the retention times depend on the length of the carbon chain and, hence, their relative hydrophobicity. For example, the retention times increase with carbon chain length. If acetonitrile is added to the system as an organic modifier, this becomes adsorbed on the resin surface and competes with the lipophilic ions in the adsorption equilibria. The blocking of adsorption sites on the surface of the resin leads, for both surface active and surface inactive ions, to a reduction in retention times.

5.2 SUPPRESSORS FOR ION PAIR CHROMATOGRAPHY

In order to chemically reduce the background conductivity of the eluant so that sensitive conductivity measurement is possible, conventional ASC-1 and CSC-1 suppressor columns have been employed. The membrane suppressors developed for HPIC could not be used for RPIPC due to the incompatibility of these suppressors with the high concentrations of organic solvents in the mobile phase. The disadvantages of the conventional suppressor columns have been discussed in Chapter 3, Section 3.2, "Suppressor Systems in Anion Chromatography." It was to circumvent these disadvantages that special cation exchange membrane suppressors, compatible with organic solvents, were developed. As described in Chapter 4, Section 4.3, this type of membrane also has a high permeability with respect to the quaternary ammonium bases employed for the analysis of anions by ion pair chromatography. Since dilute sulphuric acid is used as regenerating medium, the tetra-alkyl ammonium ions are exchanged for protons, from the regenerating medium. Analogous to the processes described in Chapter 3, Section 3.2, the reaction product is water. A concentration gradient results that is the driving force for the diffusion of protons through the wall of the membrane. The anions being analyzed are, thus, converted into their acid forms, allowing their sensitive detection via conductivity.

The analysis of cations by ion pair chromatography can be carried out using the membrane suppressor described in Chapter 3, Section 3.3 ("Suppressors for Caton Exchange Chromatography"), since these are sufficiently stable with respect to organic solvents. The introduction of these membrane suppressors eliminated the need for periodic regeneration of the suppressor columns.

TYPICAL OPERATING CONDITIONS

The regeneration of traditional suppressor columns can be achieved as described in Chapter 3, Section 3.2, "Typical Operating Conditions."

For AFS-2 membrane suppressors, a 0.0125 mol/dm^3 sulphuric acid solution is recommended as regenerating medium. If the concentration of tetra-alkyl ammonium hydroxide in the eluant is ca. 0.002 mol/dm^3 and the flow rate is ca. 1 cm^3/min, a background conductivity between 3 and 5 μS/cm can be achieved.

Since the AFS-2 membrane suppressors can be used for both ion exchange and ion pair chromatography, it is recommended that they be conditioned overnight with the appropriate regenerating medium between use for the different techniques.

5.3 EXPERIMENTAL PARAMETERS DETERMINING RETENTION TIMES

Ion pair chromatographic separations on an MPIC-phase are controlled by the following parameters:

- The nature of the lipophilic ions added to the mobile phase.
- Concentration of the lipophilic ions in the mobile phase.
- The nature of the organic modifier.
- Concentration of the organic modifier in the mobile phase.
- The nature and concentration of the inorganic modifier.
- pH of the eluant.
- Column temperature.

The effects of the individual parameters on the retention process are discussed below.

NATURE AND CONCENTRATION OF THE LIPOPHILIC IONS IN THE MOBILE PHASE

For the separation of anions, quaternary ammonium bases are the most suitable lipophilic ions. For the separation of cations, longchain, alkyl sulphonic acids are more suitable. Detection by suppressed conductivity requires that these reagents must be employed in their hydroxide or free acid forms, respectively. If the salts of these reagents are employed, conductivity detection becomes impossible due to the resulting high background conductivity. For the same reason, the reagents used for such analyses should be of the highest available purity.

The choice of the lipophilic ion depends solely on the degree of hydrophobicity of the ions being investigated. For separating strongly hydrophilic ions, more hydrophobic reagents are required. Whereas, for separating ions with long alkyl chains, reagents with strong hydrophilic properties are necessary. A review of the more common reagents used, in order of their hydrophobicity, is given in Table 5-1.

TABLE 5-1: Common reagents used for ion pair chromatography on MPIC-phases in order of increasing hydrophobicity.

ANALYSIS OF ANIONS	Hydrophobicity ↓	ANALYSIS OF CATIONS
Ammonium hydroxide		Hydrochloric acid, perchloric acid
Tetramethyl ammonium hydroxide		Hexane sulphonic acid
Tetrapropyl ammonium hydroxide		Heptane sulphonic acid
Tetrabutyl ammonium hydroxide		Octane sulphonic acid

It is, of course, possible to mix two reagents to form a mixed eluant for more complex separation problems, but this is not usually necessary.

The required concentration of the lipophilic ion is limited by two factors:

- The retention of the ions being analyzed increases with increasing lipophilic ion concentration.
- More concentrated eluant solutions lead to greater background conductivities.

For these reasons, the useful concentrations of the compounds listed in Table 5-1 generally lie between $5 \cdot 10^{-4}$ and $5 \cdot 10^{-3}$ mol/dm^3. In many cases, a concentration of $2 \cdot 10^{-3}$ mol/dm^3 has proved most suitable.

NATURE AND CONCENTRATION OF THE ORGANIC MODIFIER

Organic modifiers, such as acetonitrile or methanol, serve to increase the rate at which ions elute from the column. These compounds are adsorbed onto the resin surface and, thus, compete with the lipophilic ions in the adsorption equilibria. Hence, the analysis of the majority of ions can be achieved by varying the concentrations and nature of the lipophilic reagent and the modifier. Using the example of a mixture of inorganic anions (F^-, Cl^-, Br^-, NO_2^-, NO_3^-), the chromatogram given in Figure 5-3a shows that a separation with pure ammonium hydroxide solutions is not possible. If however, tetrapropyl ammonium hydroxide (which is more hydrophobic) is used, the individual ions are more strongly retained by the stationary phase (see Figure 5-3b). With tetrabutyl ammonium hydroxide (Figure 5-3c), the retention times become so long that only F^- elutes within an acceptable analysis time. If one adds 5 vol.-% acetonitrile to the tetrabutyl ammonium hydroxide eluant, one obtains the chromatogram shown in Figure 5-3d. Figure 5-3d shows a baseline separation of all signals with an analysis time of some 25 minutes. Furthermore, if the modifier concentration is increased to 8 vol.-%, the analysis time is further reduced without loss of resolution. When analyzing highly polarizable anions, such as I^- and SCN^-, it is generally necessary to increase the concentration of modifier.

The strongly hydrophobic longchain alkyl sulphonates require a hydrophilic reagent, such as ammonium hydroxide, for their separation. Figure 5-4a shows the chromatogram obtained from a mixture of pentane and hexane sulphonic acids using ammonium hydroxide as eluant. Although this eluant is exceedingly hydrophilic, the time required for the analysis is unacceptably long. If however, some 5 vol.-% acetonitrile is added to the eluant (Figure 5-4b), even heptane sulphonic acid elutes within 15 minutes. If the modifier concentration is further increased to 10 vol.-%, octane sulphonic acid can also be eluted (Figure 5-4c); but the resolution of the other three analytes becomes unsatisfactory. The transition to a strongly hydrophobic reagent, such as tetrapropyl ammonium hydroxide, with a correspondingly higher concentration of modifier (some 20 vol.-% acetonitrile) results in better resolution between the individual analytes, as can be seen in Figure 5-4d. The resolution, as can be seen in Figure 5-4e, is not considerably improved by using tetrabutyl ammonium hydroxide and an even higher concentration (28 vol.-%) of acetonitrile.

These two examples show that there are often several sets of parameters that will satisfactorily separate a particular compound. It is the nature of the matrix that determines the most suitable conditions for an analysis.

The optimization of a separation of *aromatic* sulphonates can be achieved in an analogous manner, although the aromatic compounds are more strongly retarded than their aliphatic homologues.

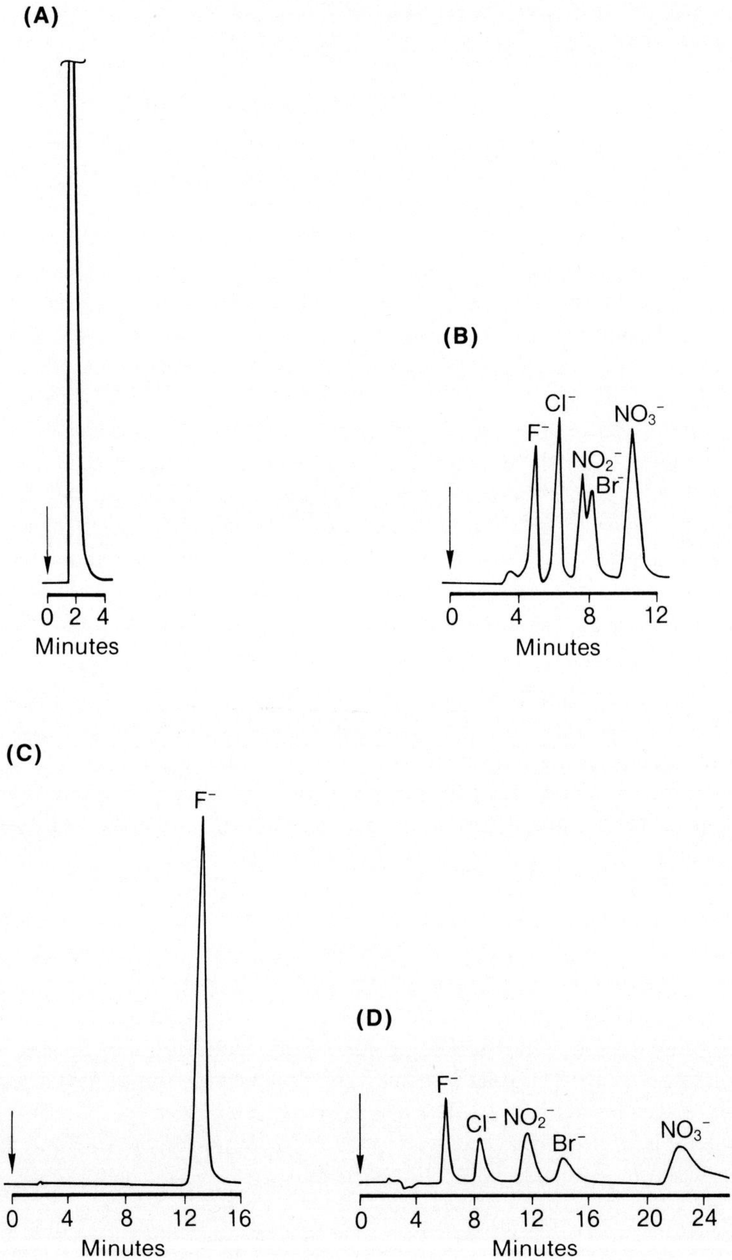

FIGURE 5-3.
The dependence of the separation of inorganic mineral acids on the nature of the lipophilic ions and the concentration of organic modifier. **Column:** MPIC-NS1. **Detection:** Suppressed conductivity. **Eluant:** (a) 0.002 mol/dm^3 NH_4OH (b) 0.002 mol/dm^3 TPAOH (c) 0.002 mol/dm^3 TBAOH (d) 0.002 mol/dm^3 TBAOH + 5 vol.-% acetonitrile.

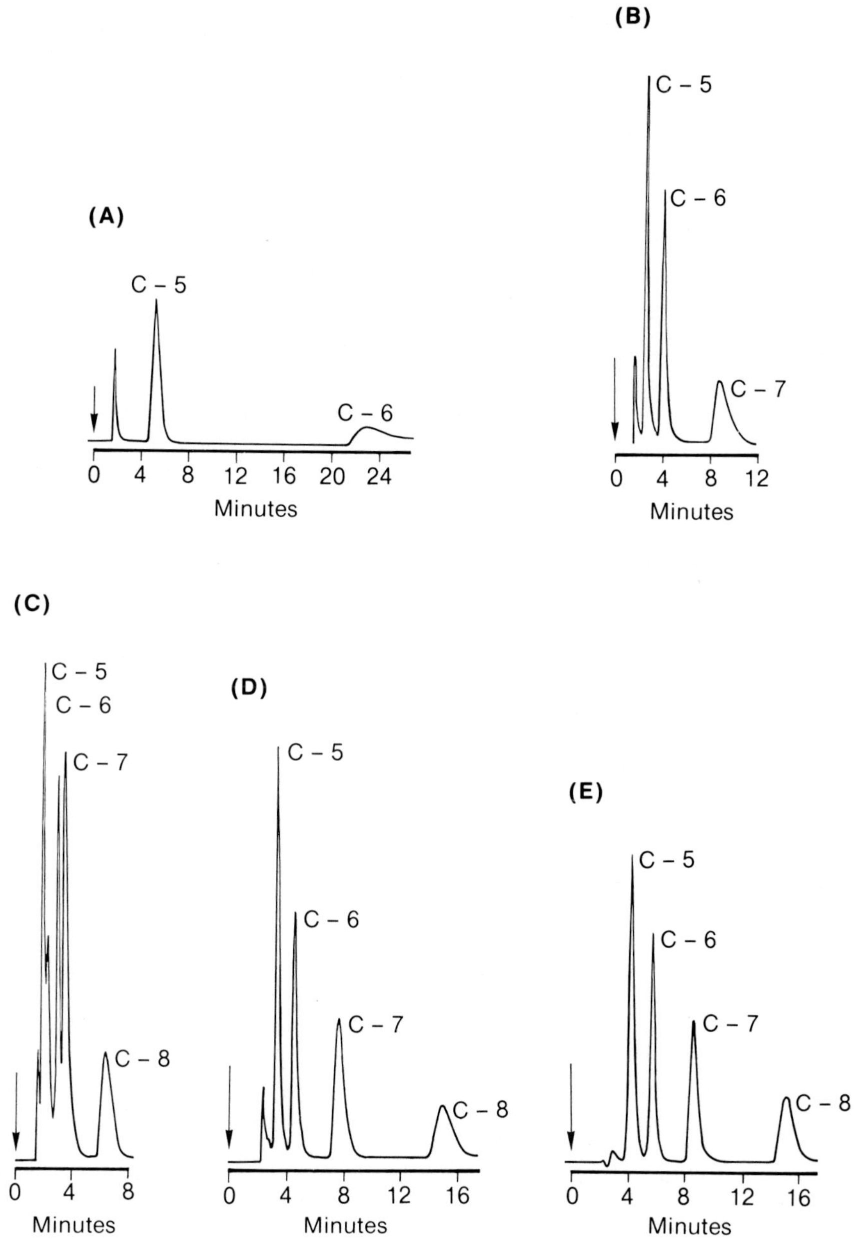

FIGURE 5-4.
Dependence of the separation of longchain alkyl sulphonates on the nature of the lipophilic ions and the concentration of organic modifier. **Column:** MPIC-NS1. **Flow rate:** 1 cm^3/min. **Detection:** Suppresssed conductivity. **Eluant:** (a) 0.01 mol/dm^3 NH_4OH (b) 0.01 mol/dm^3 NH_4OH + 5 vol.-% acetonitrile (c) 0.01 mol/dm^3 NH_4OH + 10 vol.-% acetonitrile (d) 0.002 mol/dm^3 TPAOH + 20 vol.-% acetonitrile (e) 0.002 mol/dm^3 TBAOH + 20 vol.-% acetonitrile.

As well as acetonitrile, other water miscible organic solvents can be used as modifiers. In this respect, methanol is particularly important since, due to the formation of hydrogen bonds, it allows the specificity of the system to be controlled. If methanol is substituted for acetonitrile, then the concentration of the modifier must be increased in order to achieve similar results. This effect can be demonstrated by the example of a separation of longchain aliphatic sulphonic acids. Compared to Figure 5-4b, Figure 5-5 shows a chromatogram obtained using methanol instead of acetonitrile as the modifier. It can be seen that the retention times of the alkyl sulphonates are considerably increased using the same concentration of methanol.

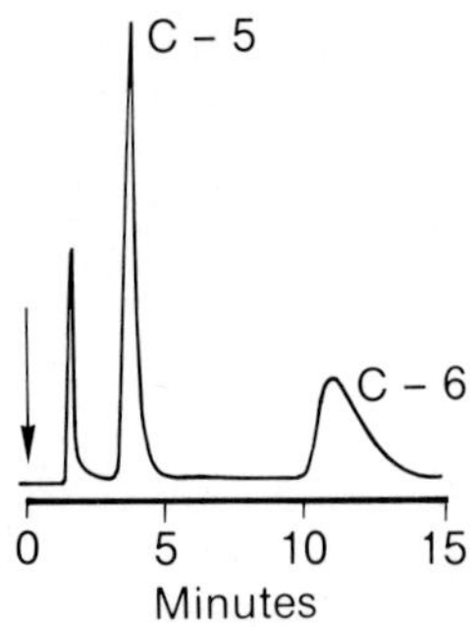

FIGURE 5-5.
The separation of longchain alkyl sulphonates with methanol as organic modifier. **Column:** MPIC-NS1. **Eluant:** 0.01 mol/dm^3 NH_4OH + 5 vol.-% MeOH. **Flow rate:** 1 cm^3/min. **Detection:** Suppressed conductivity.

One disadvantage of using methanol rather than acetonitrile as the modifier is its greater viscosity which leads to considerably increased column backpressures.

INORGANIC MODIFIERS

Inorganic modifiers, such as sodium carbonate, are added to the mobile phase in order to decrease the retention of bi- and multifunctional ions and, also, to improve the peak shape. If phosphate and sulphate are chromatographed as their divalent ions under the same conditions used to obtain the chromatogram shown in Figure 5-3d, long analysis times are required (ca. 32 min for sulphate) and the peaks are very broad. Adding 0.001 mol/dm^3 sodium carbonate to the eluant, decreases the retention times of the divalent ions and improves their peak form without noticeably affecting the retention times of the monovalent ions. The addition of small concentrations of carbonate to the eluant has a particularly large effect on the way in which hexacyanoferrate ions $[Fe(CN)_6]^{3-}$ and $[Fe(CN)_6]^{4-}$ elute. An optimum separation of these ions can be obtained with an eluant containing tetrabutyl ammonium hydroxide, acetonitrile and $2 \cdot 10^{-4}$ mol/dm^3 sodium carbonate (see Figure 5-6). If the amount of acetonitrile is reduced and that of sodium carbonate increased, the resolution between the two analytes is drastically reduced, although the peak form is further improved.

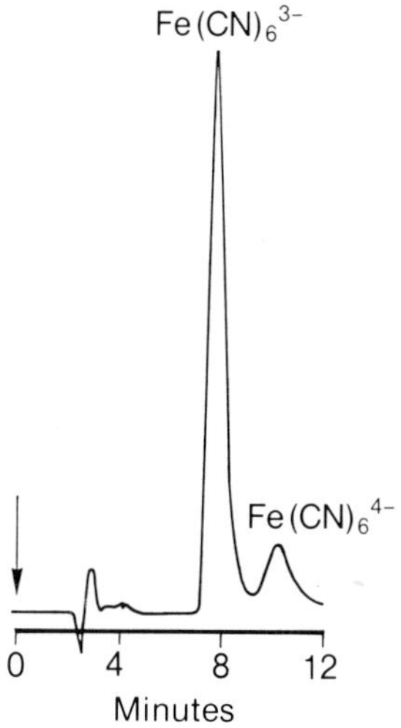

FIGURE 5-6.
The separation of hexacyanoferrate ions. **Column:** MPIC-NS1. **Eluant:** 0.002 mol/dm^3 TBAOH + 0.0002 mol/dm^3 Na_2CO_3 + 35 vol.-% acetonitrile. **Flow rate:** 1 cm^3/min. **Detection:** Suppressed conductivity.

Sodium carbonate as an inorganic modifier is also the key to the separation of sulfosuccinic acid and its mono- and diesters. The Na_2CO_3 concentration in the eluant determines the quality of the separation.

From a mechanistic point of view, the action of sodium carbonate as modifier is not yet clearly understood. It is presumed that it interferes with the adsorption equilibria and competes with the solute and lipophilic ions for adsorption sites on the surface of the stationary phase.

pH AND TEMPERATURE EFFECTS

For the analysis of the multivalent ions, it is often advantageous to adjust the pH of the eluant (caused by the lipophilic ions) by adding an appropriate acid or base. Since the retention time of multivalent ions increases with increasing degree of dissociation, the pH can be used to control retention behavior. This effect will be demonstrated using the analysis of thioglycolic acid as an example. This analysis can be achieved using an eluant containing tetrabutyl ammonium cations as lipophilic reagent and an appropriate amount of acetonitrile as modifier. Under these chromatographic conditions (pH ca. 10), no signal is obtained from thioglycolic acid. However, by decreasing the pH of the eluant with the aid of boric acid to between 7 and 7.5, the chromatogram shown in Figure 5-7 can be obtained. Boric acid is particularly useful for adjusting the pH of the eluant since, due to its low degree of dissociation, it does not lead to large increases in the background conductivity. For increasing the pH of the eluant, NaOH can be used since this, also, affects the background conductivity only slightly.

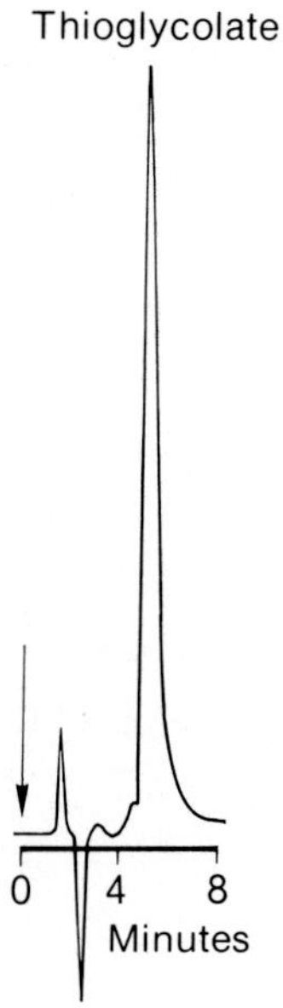

FIGURE 5-7.
Separation of thioglycolic acid. **Column:** MPIC-NS1. **Eluant:** 0.002 mol/dm^3 TBAOH + 20 vol.-% acetonitrile (pH 7–7.5, adjusted with H_3BO_3). **Flow rate:** 1 cm^3/min. **Detection:** Suppressed conductivity.

As well as controlling the degree of dissociation of the species being analyzed, the pH of the medium can be adjusted to avoid side reactions which would detract from the quality of the analysis. This is particularly true in the case of analyzing mercaptans converted into disulfides in alkaline media.

In contrast to pH, the column temperature is only rarely used for optimizing a separation. Up to a point, the elution times can be reduced by increasing the column temperature. Generally, the viscosity of the eluant will decrease and the chromatographic efficiency will increase with increasing temperature.

On the other hand, for mechanistic studies, and in particular for ion pair chromatography, the dependence of retention on temperature is an essential parameter. For ion pair chromatography, the separation mechanism is dominated by adsorption processes and these are very temperature dependent.

5.4 ANALYSIS OF SURFACE-INACTIVE IONS

For the analysis of *anions*, ion pair chromatography is important as a reference method for ion exchange chromatography. Should it be suspected that a peak results from the coelutions of two ions using one of these methods, then this peak can be re-examined using the other method since two substances will not have identical retention times under the totally different chromatographic conditions.

For example, nitrate and chlorate cannot be separated on a latex anion exchange column leading to an erroneous interpretation of the analytical results. These two ions can be differentiated by making use of their different adsorption characteristics. Since chlorate adsorbs light only in the vacuum UV, nitrate can be detected in the presence of chlorate at 215 nm. If these two substances are both present in a sample, then a quantitative analysis is only possible by differential measurements, presenting a number of associated problems. In contrast, using ion pair chromatography and an eluant containing tetrabutyl ammonium hydroxide, two discrete peaks can be obtained which become baseline resolved when the eluant contains some 9 vol.-% acetonitrile (see Figure 5-8). If one accepts the hypothesis that the stationary phase of an ion pair chromatographic column is converted into a dynamic exchange material by the addition of lipophilic ions, the difference in the selectivity of these columns—compared to conventional anion exchange resins—can only be explained in terms of the differences in the exchange functions.

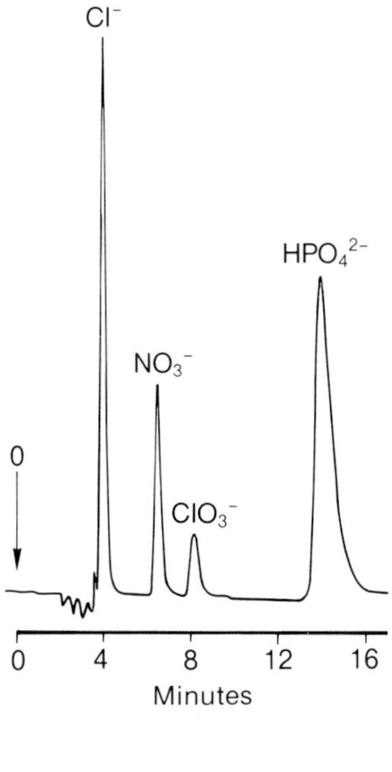

FIGURE 5-8.
The separation of nitrate and chlorate using ion pair chromatography. **Column:** MPIC-NS1. **Eluant:** 0.002 mol/dm^3 TBAOH + 9.5 vol.-% acetonitrile. **Flow rate:** 1 cm^3/min. **Detection:** Suppressed conductivity.

An additional example of such phenomena is the separation of nitrite and benzoate which cannot be achieved using anion exchange methods. Due to the interactions between the benzene ring of the benzoate anion and the polystyrene-divinyl benzene resin, the benzoate anion has a very different retention behavior to that of nitrite under ion pair chromatographic conditions. As can be seen from Figure 5-9, benzoate is more strongly retained on the stationary phase than is nitrite. These ions can be easily separated. Impressively, such samples show the importance of ion pair chromatography for the separation of even simple mixtures of inorganic ions.

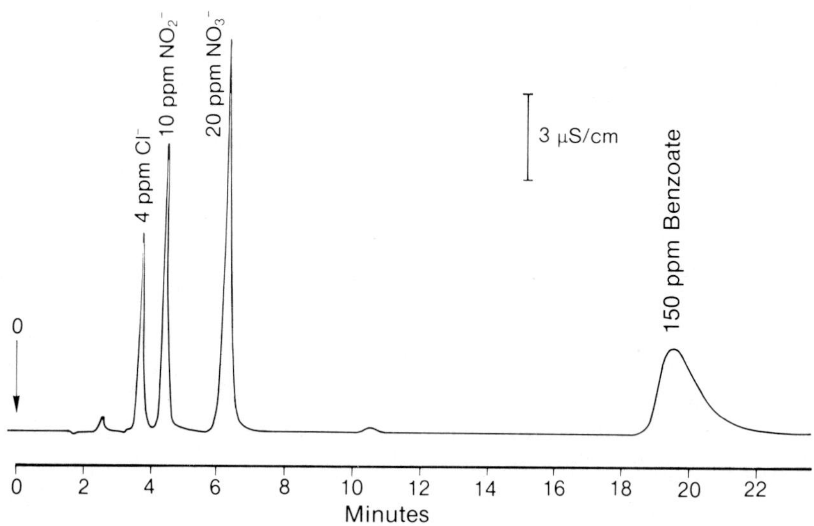

FIGURE 5-9.
The separation of nitrate and benzoate with the aid of ion pair chromatography. **Column:** MPIC-NS1. **Eluant:** 0.002 mol/dm^3 TBAOH + 8 vol.-% acetonitrile. **Flow rate:** 1 cm^3/min. **Detection:** Suppressed conductivity.

The analysis of metal-cyanide complexes is another application of ion chromatography in the environmental and plating industry. This is interesting since, prior to the introduction of ion pair chromatography, environmentally important cyanides could only be analyzed chromatographically in their free ion form. Ion pair chromatography allows the analysis of the very stable cyanide complexes of iron, gold, and cobalt.

In addition, from the nature of the cyanide complex, the oxidation state of the metal can be identified. Using the examples of iron and gold cyanide complexes, Figures 5-10 and 5-11 show that the Fe^{2+}/Fe^{3+} and Au^{+}/Au^{3+} complexes exhibit different retention behavior. Since the hexacyanoferrate complexes are multivalent, the peaks—and, in particular, the Fe (II) complex—tend to be rather broad. It is important for the separation of the two analytes that elute in proportion to their charge densities, that a small amount of sodium carbonate be added to the mobile phase. For the monovalent gold complexes, the different coordination numbers, and the resulting geometries, are responsible for the different retention times.

In terms of *cation* analysis, ion pair chromatography is used especially for the separation of amines. Whereas for shortchain (C_3-C_4) amines, cation exchange techniques—using surface sulphonated resins—can also be used, ion pair chromatographic techniques have been developed for the separation of structural

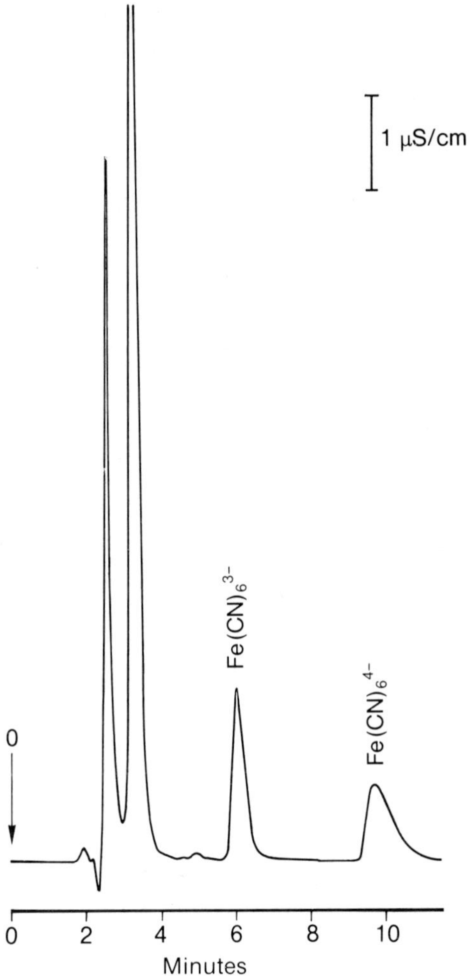

FIGURE 5-10.
The separation of iron (II) and iron (III) as their cyanide complexes. **Column:** MPIC-NS1. **Eluant:** 0.002 mol/dm^3 TBAOH + 0.0002 mol/dm^3 Na_2CO_3 + 35 vol.-% acetonitrile. **Flow rate:** 1 cm^3/min. **Detection:** Suppressed conductivity.

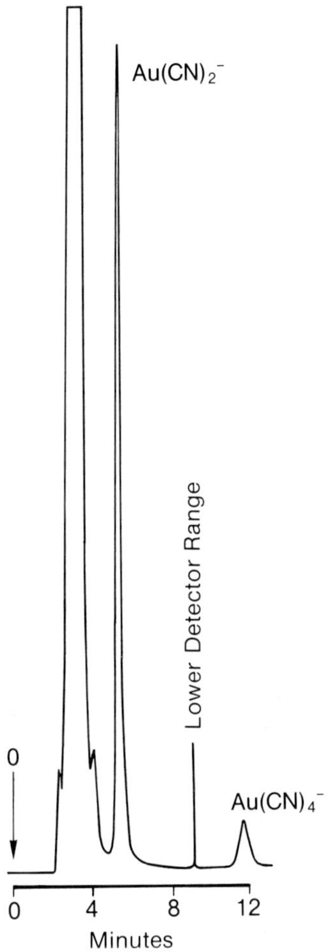

FIGURE 5-11.
The separation of gold (I) and gold (III) as their cyanide complexes. **Column:** MPIC-NS1. **Eluant:** 0.002 mol/dm^3 TBAOH + 0.0002 mol/dm^3 Na_2CO_3 + 40 vol.-% acetonitrile. **Flow rate:** 1 cm^3/min. **Detection:** Suppressed conductivity.

isomers, alcoholic amines, and small aromatic amines. As an example, Figure 5-12 shows the chromatogram of the separation of dimethyl and monomethyl amine, using an eluant containing octane sulphonic acid as lipophilic reagent. The somewhat less hydrophobic hexane sulphonic acid is used in conjunction with acetonitrile as a modifier for the separation of alcoholic amines. From Figure 5-13, it can be seen that a simultaneous analysis of monomethyl amines and the ammonium ion is possible. For all these analyses, conductivity detection can be employed and the background conductivity reduced with the aid of a modern membrane suppressor.

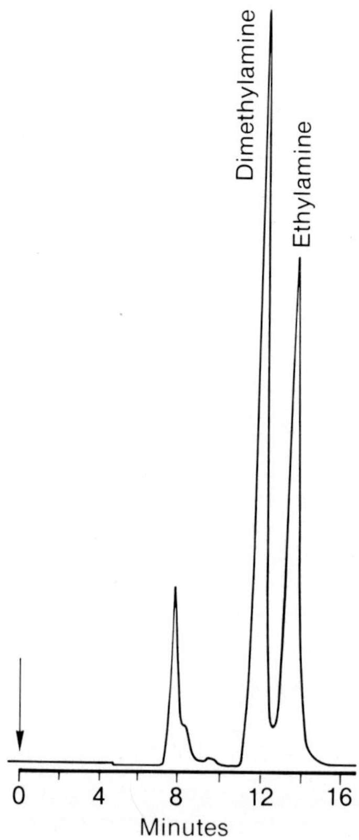

FIGURE 5-12.
The separation of dimethyl and ethyl amines with the aid of ion pair chromatography. **Column:** MPIC-NS1. **Eluant:** 0.002 mol/dm^3 octane sulphonic acid + 2 vol.-% acetonitrile. **Flow rate:** 1 cm^3/min. **Detection:** Suppressed conductivity.

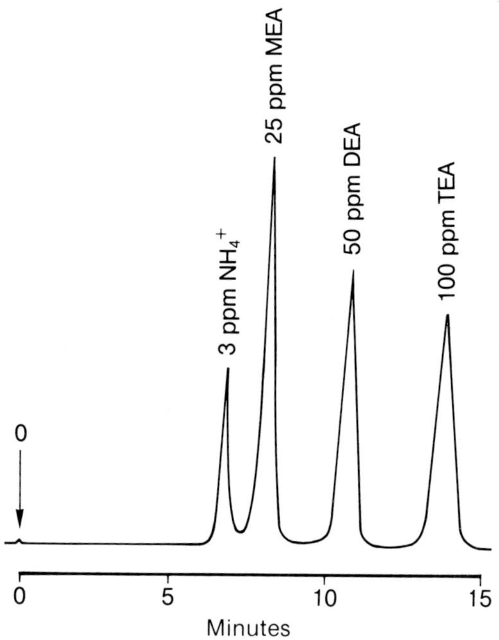

FIGURE 5-13.
The separation of ammonium and monomethanolamine. **Column:** MPIC-NS1. **Eluant:** 0.002 mol/dm^3 hexane sulphonic acid. **Flow rate:** 1 cm^3/min. **Detection:** Suppressed conductivity.

5.5 ANALYSIS OF SURFACE-ACTIVE IONS

The analysis of surface-active ions includes the determination of alkane and alkene sulphonic acids, fatty alcohol sulphates, alkyl benzene sulphonates, and α-thiofatty acid methyl esters as well as hydrotropes (tolyl-, cumyl-, and xylyl-sulphonates) which are of special interest to the soap industry. Although alkyl benzene sulphonates and the hydrotropes can be separated using HPLC methods and detected via their UV absorption, the remaining compounds mentioned above can only be determined using a conductivity detector with the aid of MPIC.

The retention times of the surface-active ions is dependent on the size of their carbon skeletons and, thus, on their hydrophobicity. The retention time increases with increasing size of the carbon skeleton. As a rule, ammonium hydroxide is used as the ion pair reagent, since the surface active agents are strongly hydrophobic. The hydrophobicity of these surface active agents is not only determined by the size of their carbon skeletons but, also, by the nature and number of their ionic and nonionic substituents. Figure 5-14 shows the different retention behavior of sulphates and sulphonates with identical carbon chain lengths. It is clear that, due to their longer retention times, alkyl sulphates are considerably more hydrophobic than the corresponding alkyl sulphonates.

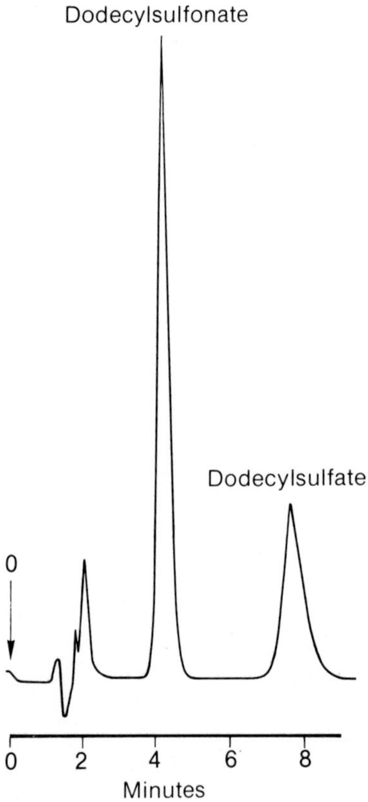

FIGURE 5-14.
The separation of dodecyl sulphate and sulphonate with the aid of ion pair chromatography. **Column:** MPIC-NS1. **Eluant:** 0.01 mol/dm^3 NH_4OH + 35 vol.-% acetonitrile. **Flow rate:** 1 cm^3/min. **Detection:** Suppressed conductivity.

Substitution of a hydrogen atom in a methylene group by an alcohol group considerably reduces the retention of the species. Thus, the retention behavior of a C_{14} hydroxy-alkyl sulphonate is almost identical with that of a C_{12} alkyl sulphonate. The influence of the actual position of the hydroxy function within the carbon skeleton on the retention time is noteworthy. Figure 5-15 shows the chromatogram from a separation of two C_{16} hydroxy-alkyl sulphonates, differing only in that the hydroxy function is sited either on the second or third carbon atom of the alkyl chain. From Figure 5-15, it can be seen that the retention time of the species decreases as the distance between the hydroxy and sulphonate functions increases.

The analysis of alkene sulphonates (rarely obtainable as pure compounds), requires considerably more effort. The various positions of the double bond within the carbon chain, as well as the possibility of cis/trans isomerization, generally lead to products being a mixture of various structural isomers. Satisfactory separations are usually possible only when the products are well defined. As an example, Figure 5-16 shows a chromatogram from a very pure C_{16} alkene sulphonate (trans-3-ene). For all alkene sulphonates, the peaks tend to exhibit

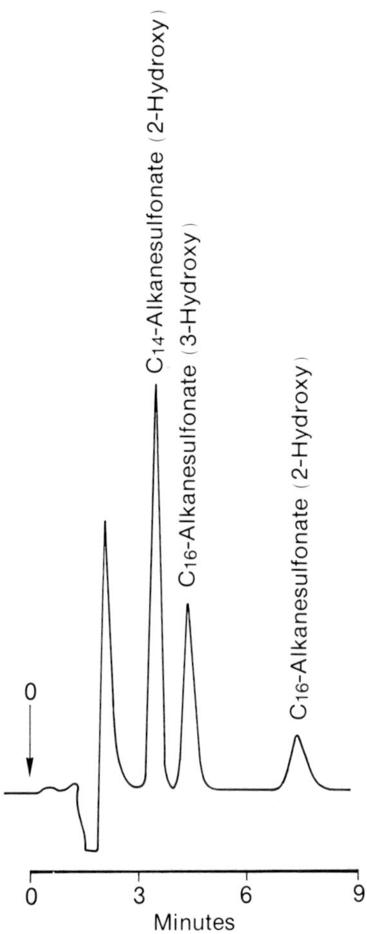

FIGURE 5-15.
The separation of two hydroxyalkane sulphonic acids whose hydroxy groups are at different sites in the alkane chain. **Column:** MPIC-NS1. **Eluant:** 0.01 mol/dm^3 NH_4OH + 37.5 vol.-% acetonitrile. **Detection:** Suppressed conductivity.

more tailing than those of alkane sulphonates with identical chain lengths. Compared with hydroxy-alkyl sulphonates, alkene sulphonates are considerably more strongly retained by the stationary phase. The difficulty of separating a technical mixture is demonstrated by Figure 5-17, showing the chromatogram of a C_{14} alkene sulphonate mixture. Using ammonium hydroxide as lipophilic reagent, only the shoulder on the major peak indicates that the mixture contains more than a single substance. By using a more hydrophobic lipophilic reagent — tetramethyl ammonium hydroxide (Figure 5-17b) — the shoulder of the major peak can be better resolved. Also, an additional compound, with a slightly shorter retention time than the major component, becomes visible. The retention times are, however, increased since the concentration of organic modifier was not increased. An even better resolution can be obtained by using tetrapropyl ammonium hydroxide as lipophilic reagent (see Figure 5-17c). The strong hydrophobicity of this reagent leads to longer retention times, but permits the identification of five components in addition to the major component. This analysis

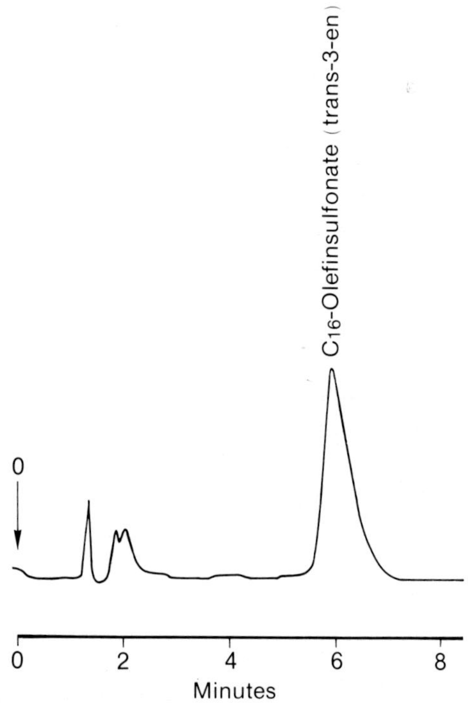

FIGURE 5-16.
The separation of a C_{16} alkene (trans-3-ene) sulphonic acid. **Column:** MPIC-NS1. **Eluant:** 0.01 mol/dm^3 NH_4OH + 40 vol.-% acetonitrile. **Flow rate:** 1 cm^3/min. **Detection:** Suppressed conductivity.

was also carried out without changing the proportion of organic modifier in the eluant. These three examples show the limits of this technique for separating technical mixtures of alkene sulphonates, despite the fact that the major component is separable. Considerable improvements can only be achieved by using resins having smaller particle sizes.

Among the most important anionic surfactants in the soap industry are the linear alkyl benzene sulphonates (LAS):

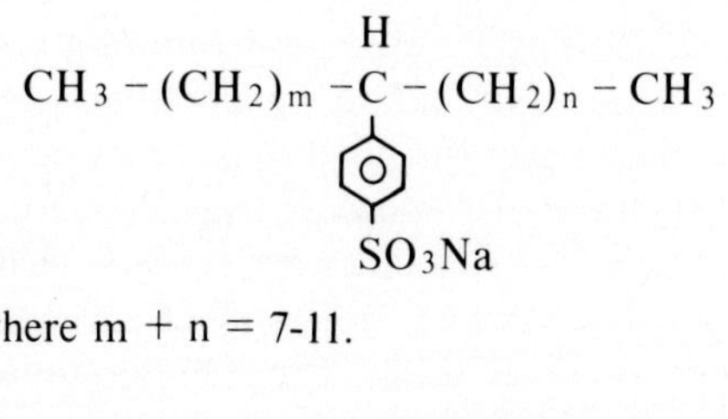

(LAS)

where m + n = 7-11.

The aromatic ring of the LAS may be bound anywhere along the alkyl chain and the industrial products are usually statistical mixtures. Furthermore, since the alkyl chains vary in length, a satisfactory separation cannot be expected. Nevertheless, by comparing the "fingerprint" of a product, valuable information for raw material control can be obtained.

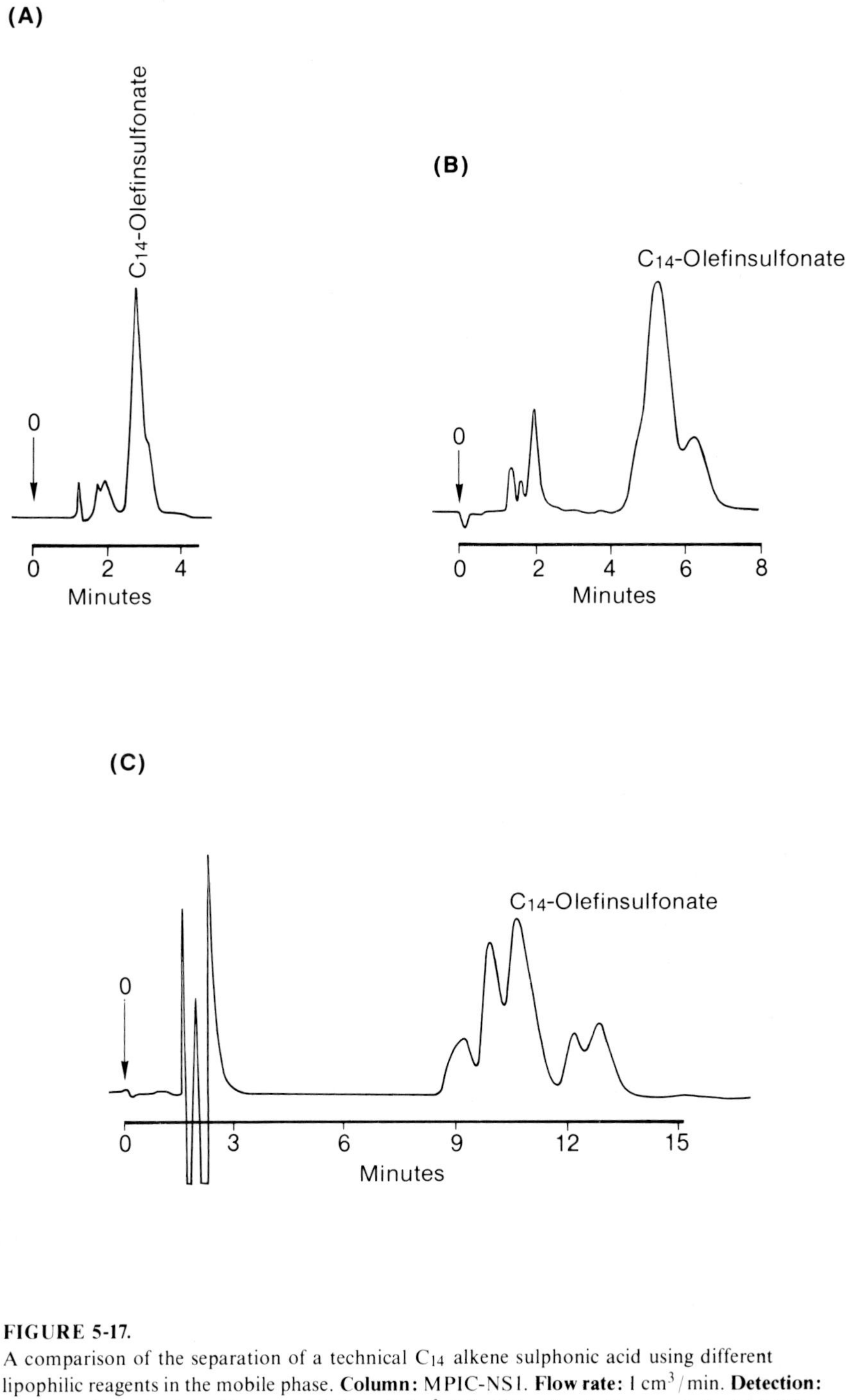

FIGURE 5-17.
A comparison of the separation of a technical C_{14} alkene sulphonic acid using different lipophilic reagents in the mobile phase. **Column:** MPIC-NS1. **Flow rate:** 1 cm^3/min. **Detection:** Suppressed conductivity. **Eluant:** (a) 0.01 mol/dm^3 NH_4OH + 40 vol.-% acetonitrile (b) 0.002 mol/dm^3 TMAOH + 40 vol.-% acetonitrile (c) 0.002 mol/dm^3 TPAOH + 40 vol.-% acetonitrile.

The analysis of hydrotropic compounds — such as tolyl-, xylyl-, and cumyl-sulphonates — can be achieved very easily in contrast to the difficult separations discussed above. These compounds elute in order of increasing alkyl substitution. Figure 5-18 shows a typical chromatogram. Aromatic sulphonates, which always elute before their straight chain analogues, have the advantage that they can be detected both by conductivity and UV absorption techniques. The choice of the appropriate technique depends, essentially, on the nature of the matrix.

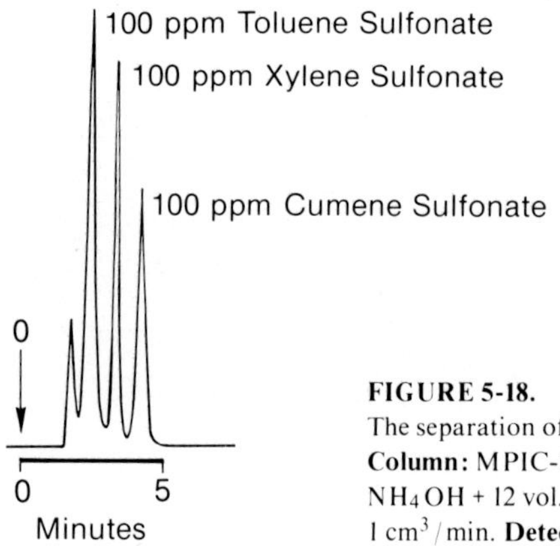

FIGURE 5-18.
The separation of hydrotropic compounds. **Column:** MPIC-NS1. **Eluant:** 0.01 mol/dm^3 NH_4OH + 12 vol.-% acetonitrile. **Flow rate:** 1 cm^3/min. **Detection:** Suppressed conductivity.

Ion chromatography of *surface-active cations* is suitable for the analysis of quaternary ammonium and phosphonium compounds. Among the most important compounds are:

- The dialkyl-dimethyl ammonium chlorides

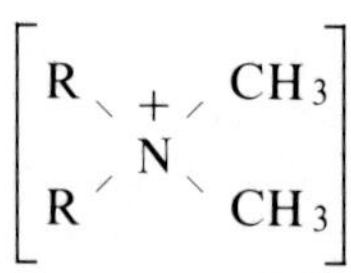

Cl^- $R = C_{16} - C_{18}$

- The imidazolinium salts

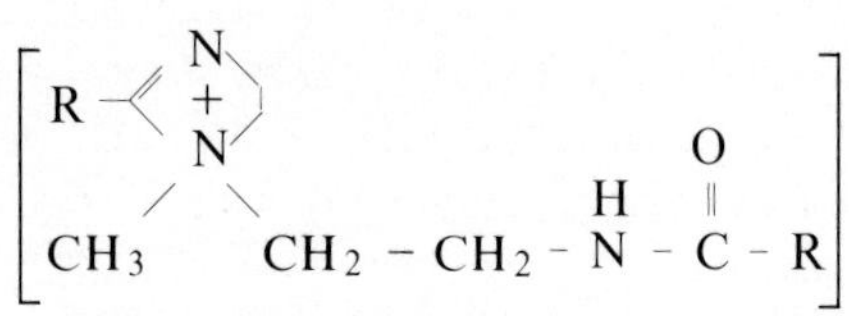

$CH_3 - O - SO_3^-$

$R = C_{16} - C_{18}$

- And the alkyl-dimethyl-benzyl ammonium chlorides

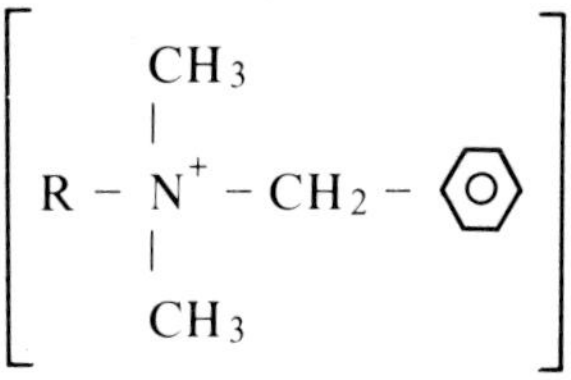

Cl^- $R = C_8 - C_{18}$

as well as cetyl pyridinium chloride; all of which are used in the soap industry. In the past, cationic surfactants of these types were not industrially important; but, in recent years, they have become increasingly more widely used. Thus, the liquid chromatographic separation of such compounds has rarely been described in existing literature. Initial experiments with cetyl pyridinium chloride (Figure 5-19) do, however, suggest that ion pair chromatography is a suitable alternative to the more time consuming and labor intensive wet chemical analytical techniques. Some of the cationic surfactants have unusually long alkyl chains. It is, thus, essential to use the perchlorate anion, which has a low hydrophobicity, as ion pair reagent. In order to reduce the analysis time to an acceptable level, a high percentage of organic modifier is usually added.

50 ppm Cetylpyridiniumchloride

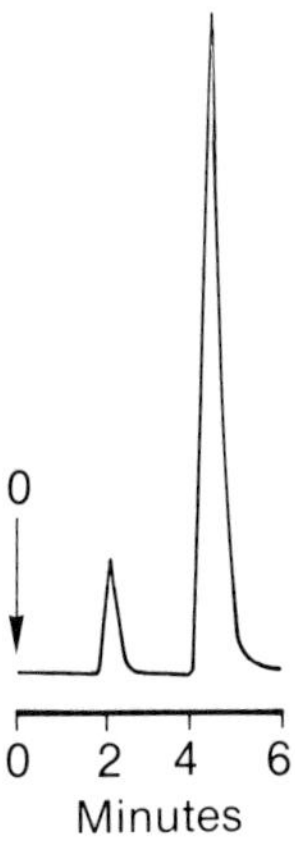

FIGURE 5-19.
Ion pair chromatographic separation of the cetyl pyridinium cation. **Column:** MPIC-NS1. **Eluant:** 0.005 mol/dm^3 $HClO_4$ + 70 vol.-% acetonitrile. **Flow rate:** 1 cm^3/min. **Detection:** Suppressed conductivity.

An example of the analysis of quaternary phosphonium salts is the chromatographic separation of triphenyl mono (β-ionylidene-ethylene) phosphonium chloride as shown in Figure 5-20. This compound, a precursor of vitamin A, has a strongly hydrophobic character and, thus, the chromatographic conditions are essentially the same as those used to obtain Figure 5-19. Since phosphonium compounds of this type contain double bonds, their detection via UV spectroscopy is considerably more sensitive than via their conductivity with a suppressor system.

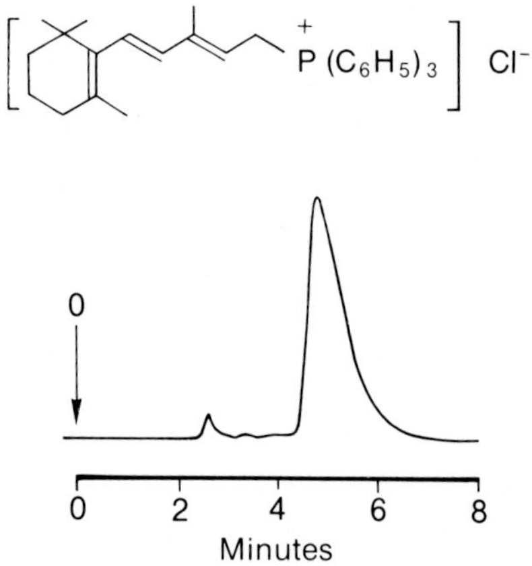

FIGURE 5-20.
Ion pair chromatographic separation of a quaternary phosphonium salt. **Column:** MPIC-NS1. **Eluant:** 0.005 mol/dm^3 $HClO_4$ + 95 vol.-% methanol. **Flow rate:** 1 cm^3/min. **Detection:** UV absorption at 280 nm.

In contrast to the silica-based stationary phases used for RPIPC, the MPIC-phases, based on polystyrene-divinyl benzene, allow the separation of compounds containing more than one quaternary ammonium group in a single run. The elution sequence depends on the nature of the other functional groups contained in the molecules.

LITERATURE

1. D.P. Wittmer, N.O. Nuessle, and W.G. Haney jr., Anal. Chem. **47,** 1422 (1975).
2. S.P. Sood, L.E. Sartori, D.P. Wittmer, and W.G. Haney, Anal. Chem. **48,** 796 (1976).
3. "Paired Ion Chromatography, an Alternative to Ion Exchange," Waters Associates, Milford, Mass. 1975.
4. J.H. Knox and G.R. Laird, J. Chromatogr. **122,** 17 (1976).
5. J.H. Knox and J. Jurand, J. Chromatogr. **125,** 89 (1976).
6. C. Horvath, W. Melander, I. Molnar, and P. Molnar, Anal. Chem. **49,** 2295 (1977).
7. C. Horvath, W. Melander, and I Molnar, J. Chromatogr. **125,** 129 (1976).
8. J.C. Kraak, K.M. Jonker, and J.F.K. Huber, J. Chromatogr. **142,** 671 (1977).
9. N.E. Hoffmann and J.C. Liao, Anal. Chem. **49,** 2231 (1977).
10. P.T. Kissinger, Anal. Chem. **49,** 883 (1977).
11. W.R. Melander and C. Horvath, J. of Chromatogr. **201,** 211 (1980).
12. J.H. Knox and J. Jurand, J. Chromatogr. **103,** 311 (1975).
13. B.A. Bidlingmeyer, S.N. Deming, W.P. Price jr., B. Sachok, and M. Petrusek, J. Chromatogr. **186,** 419 (1979).
14. B.A. Bidlingmeyer, J. Chrom. Science **18,** 525 (1980).
15. C. Pohl, "Mobile Phase Ion Chromatography (MPIC). Theory and Separation," Dionex Dept. of Research and Development, IC-Exchange Nr. 2 (1982).

6 Modes of Detection

6.1 CONDUCTIVITY DETECTORS

Detection via conductivity is a universal method for detecting ionic compounds and is the most common method used in ion chromatography. For this reason, the basic theory of this method will be summarized below.

THE ELECTRICAL CONDUCTIVITY OF AN ELECTROLYTE SOLUTION [1]

Ohms law is valid for the electrical conductivity of an electrolyte solution; i.e.,

$$R = \frac{V}{I} \tag{97}$$

V: Voltage.
I: Current.
R: Resistance.

Since the resistance is dependent on the nature of the conductor—a material specific quantity—the resistivity, ρ, is defined as:

$$\rho = \frac{A \cdot R}{\ell} \tag{98}$$

A: The cross sectional area of the conductor.
ℓ: The length.

The inverse of the resistivity is given by the conductivity, γ, and has the units $S \cdot cm^{-1}$. Thus,

$$\gamma = \frac{\ell}{\rho} \tag{99}$$

The conductivity of an electrolyte is very dependent on its concentration. In order to be able to compare one electrolyte with another, the conductivity is divided by the equivalent concentration, c_{ev}, to yield the equivalent conductivity Λ ($S\ cm^2\ val^{-1}$):

$$\Lambda = \frac{\gamma}{c_{ev}} \tag{100}$$

If two electrodes suspended in an electrolyte solution are connected to a voltage source, then an electric field of strength, E, will be generated between these electrodes. The mass transport of the cations to the negative electrode and the anions to the positive electrode within this field is called ionic migration. A measure of this migration that is independent of the strength of the field can be obtained by dividing the rate at which the ions migrate, v, by the strength of the electric field, E. This quantity is named the mobility of the individual ions, μ ($cm^2\ V^{-1}\ s^{-1}$):

$$\mu = \frac{v}{E} \tag{101}$$

If in a solution of a strong electrolyte, the concentrations of the cations and anions are given by c_+ and c_-, respectively, then:

$$c = c_+ = c_- \tag{102}$$

If an electric field is imposed on the electrolyte solution, then in a time, t, $N_A \cdot c_+ \cdot v_+ \cdot A \cdot t$ cations and $N_A \cdot c_- \cdot v_- \cdot A \cdot t$ anions travel through the cross sectional area, A. Since the ions carry an electrical charge, this movement of ions results in a transport of electrons. The quotient of the sum of the charges and the time gives the size of the electrical current, I:

$$I = I_+ + I_- \tag{103}$$

$$I = e \cdot N \cdot c_+ \cdot v_+ \cdot A + e \cdot N \cdot c_- \cdot v_- \cdot A \tag{104}$$

With $F = N \cdot e$ and Equation 101, one obtains:

$$I = F \cdot c \cdot A \cdot E\ (\mu_+ + \mu_-) \tag{105}$$

and since

$$\gamma = \frac{I}{A \cdot E} \tag{106}$$

an expression for the conductance of a strong electrolyte

$$\gamma = c \cdot F\ (\mu_+ + \mu_-) \tag{107}$$

can be obtained. For an electrolyte composed of ions carrying a charge, z, Equation 107 can be generalized in the form:

$$\gamma = c \cdot z \cdot F\ (\mu_+ + \mu_-) \tag{108}$$

If the electrolyte is not completely dissociated, then this must be taken into account. Thus,

$$\gamma = \alpha \cdot z \cdot c \cdot F\ (\mu_+ + \mu_-) \tag{109}$$

The conductance of an electrolyte solution is greater:

- The greater the concentration.

- The larger the valence of the individual ions.
- The more mobile the ions.

From Equation 100 and 108, one obtains:

$$\Lambda = \frac{\gamma}{z \cdot c} = F\,(\mu_+ + \mu_-) \qquad (110)$$

According to Kohlrausch, the quantities $F \cdot \mu_+$ and $F \cdot \mu_-$ are defined as the equivalent conductivities of the individual ions Λ_+ and Λ_-, respectively. The equivalent conductivity of an electrolyte is, thus, given by the sum:

$$\Lambda = \Lambda_+ + \Lambda_- \qquad (111)$$

The equivalent conductivity of a strong electrolyte decreases continually as its concentration increases. As the concentration decreases, the equivalent conductivity tends to a limiting value; i.e., the equivalent conductivity at infinite dilution Λ_∞ of the electrolyte being studied. This phenomenon can be explained by a decrease in the mobility of the ions in more concentrated solutions. As the concentration of the electrolyte increases, the ions come into such close contact that they are influenced by their electrostatic charges. As the time averaged distance between the ions decreases, the Coulombic interactions increase. These interionic interactions are reinforced by the decrease in the dielectric constant of the solvent in the presence of an electrolyte. These electrostatic interactions can lead to solvated ions joining together to form *ion pairs* which do not contribute to the conductivity of the solution [2].

Kohlrausch found that the equivalent conductivity of a strong electrolyte depends on its concentration according to:

$$\Lambda = \Lambda_\infty - k \cdot \sqrt{c} \qquad (112)$$

(Kohlrausch's Square Root Law)

The k term of Kohlrausch's square root law is dependent on the number of charges on the ions of the electrolyte. A plot of Λ against $\sqrt{c}$, gives a straight line whose slope is characteristic for the electrolyte type. It should, however, be noted that Equation 112 is valid only for concentrations below ca. $< 10^{-2}$ mol/dm^3.

Generally, as the temperature of a solution increases, so does its conductance. The reason for this is, in part, due to the exponential decrease in the viscosity of the solvent as the temperature increases. Increased temperature also increases the mobility of the ions.

Only the sum of the conductivities of the positive and negative ions in the solution can be derived from conductance measurements. No information about the values for the individual ions can be obtained directly. However, it is known that the equivalent conductivities of the individual ions vary considerably. Kohlrausch's law is based on the assumption that the ions move independent of one another in an infinitely dilute solution. If the equivalent conductivity of two ions are different, the ions will not make an equal contribution to the current transport. The contribution which an ion, i, makes to the current transport, is given by its transport number, t_i, defined by:

$$t_i = \frac{I_i}{I} \tag{113}$$

From Equation 105, one can derive:

$$I_+ = F \cdot c \cdot A \cdot E \cdot \mu_+ \tag{114}$$

Using these two equations, the proportion of current transported by a cation can be given, in terms of its transport number, t_+, by:

$$t_+ = \frac{I_+}{I} = \frac{\Lambda_+}{\Lambda_+ + \Lambda_-} = \frac{\Lambda_+}{\Lambda} \tag{115}$$

The corresponding equation for the anion is:

$$t_- = \frac{I_-}{I} = \frac{\Lambda_-}{\Lambda_+ + \Lambda_-} = \frac{\Lambda_-}{\Lambda} \tag{116}$$

The equivalent conductivity of most cations and anions is between 35 and 80 $S \cdot cm^2\ val^{-1}$. However, that of the hydronium ion is 350 $S \cdot cm^2\ val^{-1}$ and that of the hydroxyl ion is 198 $S \cdot cm^{-2}\ val^{-1}$ (25°C).These values are only large in aqueous solutions. Thus, a particular mechanism must be assumed for the movement of the H^+ and OH^- ions in such solutions compatible with the structure of bulk water. Hydrogen bonds, between associated water molecules, allow the protons and hydroxyl ions to be exchanged along long chains of molecules without the hydrated ions themselves actually moving. In the suppressor technique, the higher equivalent conductivity of H^+ and OH^- is utilized to convert the salts being analyzed into more strongly conducting species and to convert the eluant into less conducting species.

IONIC INTERACTIONS

For dilute electrolyte solutions, Lewis and Randall found that the mean activity coefficient of a strong electrolyte is dependent, not on the nature of the electrolyte, but rather, on the charge numbers of *all* the ions present in the solution. Thus, the properties of the individual ions in a dilute electrolyte solution do not determine the degree of ionic interaction. This observation led to the introduction of the concept of ionic strength, I:

$$I = \frac{1}{2} \sum c_i z_i^2 \tag{117}$$

The ionic strength of a solution is a very useful quantity. From it, the mean activity coefficient of an electrolyte can be easily calculated. Lewis developed the following empirical relationship:

$$\lg f_{\pm} = - A \cdot z_{+} \cdot |z_{-}| \sqrt{I} \tag{118}$$

A plot of $\lg f_{\pm}$ vs. $\sqrt{I}$ gives a straight line whose slope is a function of the charge numbers of the individual ions of the electrolyte.

According to the theory of Debye and Hückel, it is assumed that the interactions between the ions occurring in the crystal lattice do not totally disappear in solution. Thus, in solution each ion is preferentially surrounded by a sphere of ions of the opposite charge. Even though the ions are subject to Brownian motion, the time averaged picture is one of each ion being the center of an "ionic cloud" of opposite charge. Each ion within this "ionic cloud" interacts with the central ion and the total charge of the "ionic cloud" is equal to the charge on the central ion so that electroneutrality is maintained.

The distribution of the ions can, according to Debye and Hückel, be calculated with the aid of Boltzmann's energy distribution law. The use of Boltzmann's law presumes that the "ionic cloud" builds a charge space whereby the charge is greatest close to the central ion and decreases with increasing distance from its center. A number of simplifying assumptions are made about the state of the ion:

- The electrolyte is completely dissociated.
- The ionic interactions involve only electrostatic forces.
- The energy involved in the electrostatic interactions is small compared to the thermal energy of the ions.
- The ions are presumed to be spherically symmetrical, point charges (i.e., they are nonpolarizable).
- The dielectric constant of the solution is uniform and equal to the bulk value for the solvent.

As the concentration of an electrolyte solution increases, these simplifications reflect less and less the real situation because:

- Strong electrolytes form ion pairs at higher concentrations.
- At high concentrations, as well as ionic interactions, ion-molecule interactions become important. These affect the solvation state of the ions and, also, the structure of the solvent.
- At high concentrations, the electrostatic interactions restrict the thermal motion of the ions.
- As the concentration increases, ionic interactions can lead to the ions becoming polarized.
- Interactions between the ions and the solvent alter the latters dielectric constant.

According to Lewis, the constant A in Equation 118 is empirical. The Debye-Hückel Theory allows this constant to be given a physical meaning and, indeed, to be calculated. Thus, from the Debye-Hückel Theory:

$$\lg f = -A \cdot z^2 \cdot \sqrt{I} \qquad (119)$$

This allows the following conclusions to be made:

- At finite ionic strengths, lg f is always negative; i.e., the activity coefficient is always less than 1.
- The logarithm of the activity coefficient decreases from its limiting value ($\lg f = 0$, $I = 0$) in proportion to the square root of the ionic strength.
- Ions with greater charge numbers will behave less ideally than singly charged ions.
- The activity coefficient is very dependent on the temperature and the dielectric constant of the solution.

Because of the limited validity of the simplifications on which the Debye-Hückel Theory is based, Equation 119 represents a limiting law. Thus, Equation 119 describes the behavior of 1-1 electrolytes in aqueous media up to concentrations of ca. 10^{-2} mol/dm^3.

From the Debye-Hückel Theory, it is also possible to calculate the effect of the "ionic cloud" on the mobility of the ions. If an ion is moving in an electric field, then the "ionic cloud" around this ion must be continually renewed. This renewal process requires a certain time—the relaxation time. Thus, as shown schematically in Figure 6-1, the "ionic cloud" becomes asymetrical under the influence of an external electric field. The cloud is thinner in front of the ion and thicker behind it. This asymmetry leads to an electrostatic braking action and restricts the movement of the central ion. Furthermore, since the ions in the "ionic cloud" have an opposite sign to that of the central ion, the ions in this cloud—and their solvation spheres—are moving in the opposite direction to that of the central

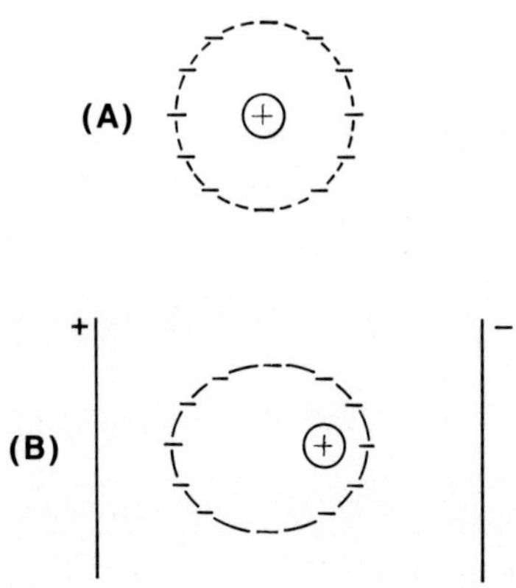

FIGURE 6-1.
Schematic representation of a central ion and its "ion cloud:" (a) without the effect of an external electric field (b) in an external electric field.

ion. Thus, the central ion is not moving through a motionless medium, but rather against a flow of ions and their solvation spheres. The effect of this contrary flow, the electrophoretic effect, is to further reduce the mobility of the central ion.

Both the relaxation and the electrophoretic effects increase with increasing electrolyte concentration and work to reduce the equivalent conductivity of the central ion. According to Onsager, a quantitative evaluation of the above effects for a 1-1 electrolyte gives:

$$\Lambda = \Lambda_\infty - \underbrace{\frac{8.2 \cdot 10^5}{\left(\varepsilon \cdot \frac{T}{K}\right)^{3/2}} \cdot \Lambda_\infty \cdot \sqrt{\frac{c}{mol \cdot dm^{-3}}}}_{\text{Relaxation}}$$

$$- \underbrace{\frac{82.48\ cm^2 \cdot S \cdot mol^{-1}}{\frac{\eta}{g \cdot cm^{-1} \cdot s^{-1}} \cdot \left(\varepsilon \cdot \frac{T}{K}\right)^{1/2}} \cdot \sqrt{\frac{c}{mol \cdot dm^{-3}}}}_{\text{Electrophoretic}}$$

$$\Lambda = \Lambda_\infty - (\text{Relaxation effect}) - (\text{Electrophoretic effect}) \tag{120}$$

It can be seen that the smaller the dielectric constant, ε, of the solvent, the greater the influence of the relaxation and electrophoretic effects on the equivalent conductivity of an ion. The electrophoretic effect is largely a measure of the viscosity of the solvent. A knowledge of the electrophoretic and relaxation constants for a solvent permits the calculation of a conductivity coefficient:

$$f_\Lambda = \frac{\Lambda}{\Lambda_\infty} = 1 - \left(A' + \frac{B}{\Lambda_\infty}\right)\sqrt{c} \tag{121}$$

6.2 UV/VIS DETECTION

In contrast to more conventional HPLC for ion chromatography, UV/Vis detection plays only a subsidiary role, although it can often be a useful addition to the other methods. Direct detection has the disadvantage that most inorganic ions do not have appropriate chromophores and only absorb below 220 nm [3]. This is corroborated by the work of Reeve [4] and Leuenberger [5] who separated inorganic ions on chemically bound cyano- and amino phases and detected the eluting ions at 210 nm. The more recent work of Williams [6] has shown the advantages to be gained from using both conductivity and UV detection in connection with a suppressor system. However, the wavelength used by Williams (λ = 195 nm) restricts the use of his system to very simple matrices and, thus, this work will probably remain of only academic interest. The possible, practical application of this technique to surface- and wastewater analyses is, in particular, limited by the large number of interfering substances absorbing above 195 nm. Furthermore, UV detectors are some ten times less sensitive than modern conductivity detectors. In contrast, direct UV detection is suitable for the analysis of nitrite, nitrate, bromide, and iodide in the presence of larger concentrations of chloride ions. Figure 6-2 demonstrates the superiority of direct UV detection for the analysis of nitrite in the presence of a 100 × excess of chloride. Determinations of this type are, for example, the analysis of body fluids such as human saliva and stomach juices.

An important application of UV/Vis detection is the photometric analysis of the column effluent after derivatization. A good example of this method is the

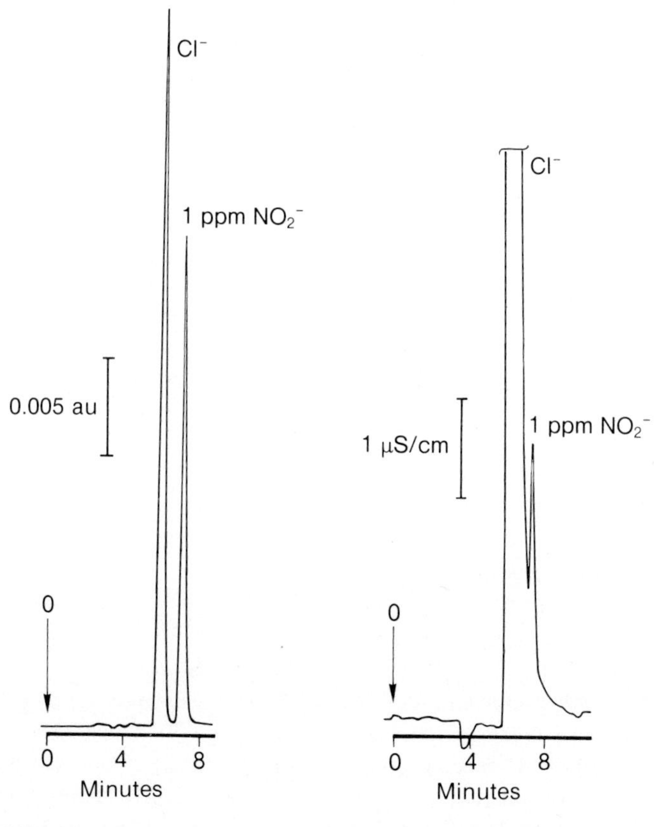

FIGURE 6-2.
A comparison of direct UV detection with detection via suppressed conductivity for the analysis of nitrite in the presence of a 100-fold excess of chloride.

analysis of transition and heavy metals after their reaction with 4-(2-pyridylazo)-resorcinal (PAR). The metal ions (separated on an HPIC column as their oxalate/citrate complexes) are mixed with PAR reagent in a postcolumn reactor and form chelate complexes that absorb at approximately 520 nm. The PAR reagent is introduced into the postcolumn reactor using constant pressure in order to maintain a pulse free flow. Initially, the postcolumn reactor was a short column of appropriate dimensions and filled with chemically inert beads. In this way, peak broadening was minimized while ensuring optimum mixing. The introduction of the new membrane reactors has considerably improved these techniques. The membrane reactors are connected to a reservoir of the appropriate reagent and are constructed from a novel semipermeable membrane. The reagent is forced, under pressure, through the outer channel of the membrane reactor and diffuses through the membrane into the inner channel where it reacts with the column effluent. This method of introducing the reagent leads to considerably less short-term noise in connection with subsequent photometric detection. Figure 6-3 shows this effect, which is particularly noticeable when making very sensitive measurements, by comparison with the conventional reagent introduction method. Since the required volume of reagent is also reduced, the dilution effect is minimized. This optimization of both parameters

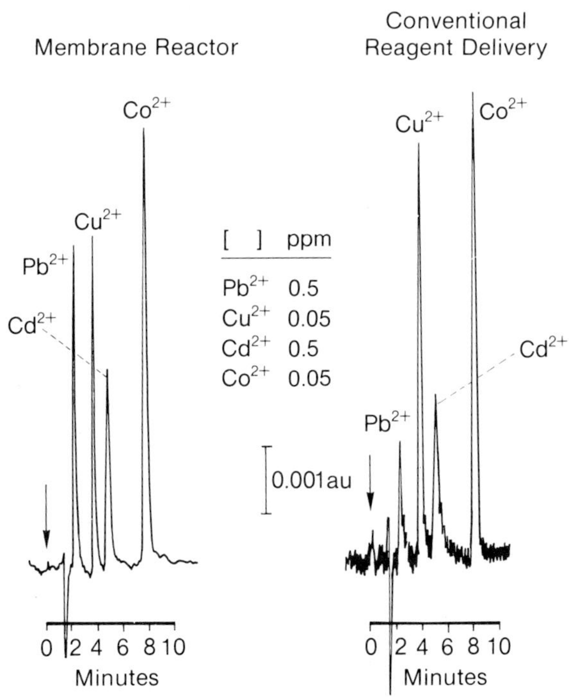

FIGURE 6-3.
A comparison of signal to noise ratios using conventional and semipermeable membrane postcolumn derivatization.

has led to a considerable increase in the available sensitivity. Postcolumn derivatization, in conjunction with photometric detection, has also been developed for the analysis of aluminum and polyvalent anions. Aluminium can be separated as $AlSO_4^+$ using a cation exchange column. It is then converted into a stable complex with the disodium salt of 4,5-dihydroxyl-1, 3-benzene disulphonic acid (Tiron) which has an absorption maximum at 313 nm. Polyvalent anions, such as polyphosphonates, chelants, and condensed phosphates, are eluted from special anion exchange columns using nitric acid as eluant. They can be derivatized using iron (III) nitrate in acid solution to form the corresponding complexes having absorption maxima around 330 nm.

Analysis of water soluble silicates is achieved by separating them using either ion pair or ion exclusion chromatography and derivatizing the eluted ions with sodium molybdate, Na_2MoO_4, in acid solution. Silicates react with molybdic acid to form a yellow hetero-poly acid, $H_4[Si(Mo_3O_{10})_4]$, having an absorption maximum at 410 nm.

A sensitive analysis of calcium and magnesium in solutions of sodium chloride ($c < 30\%$) is also possible using postcolumn derivatization [7]. After a separation on a cation exchange column, the effluate is treated with Arsenazo I [o-(1, 8-dihydroxy-3,6-disulpho-2-naphthyl azo)-benzene arsenic acid]. Arsenazo I forms complexes with both calcium and magnesium ions, but not with sodium and, thus, is a specific reagent. The complexes can be detected at 570 nm and, using appropriate concentration techniques, the sensitivity of the method is in the low ppb range.

Indirect UV analysis of the effluent is also possible. In 1982, Small, et al., [8] and Cochrane and Hillman [9] independently developed indirect photometric chromatography—IPC. This technique involves the use of a UV absorbing eluant to analyze UV transparent anions. The resulting chromatogram consists of negative peaks corresponding to the eluting ions. This, and other techniques related to ion chromatography, will be described in more detail in Chapter 10. IPC has been successfully used for the quality control of major components and is also suitable for the analysis of polyvalent anions when they are present in higher concentrations.

6.3 AMPEROMETRIC DETECTION

Detection by amperometry is used for ions whose pK values are greater than seven. Such ions form acids (or bases) as products of the suppressor reaction which are only slightly dissociated and, therefore, cannot be detected via conductivity.

Conventional amperometric detectors consist of a three-electrode measuring cell containing a working electrode, a reference electrode, and a counter electrode. The potential required for oxidation, or reduction of the species being analyzed, is applied between the working electrode and the Ag/AgCl reference electrode. The counter electrode is usually made of "glassy carbon" and serves to preserve the potential during operation. The counter electrode also hinders a flow of current at the reference electrode that would lead to the latter's destruction. If an electrochemically active substance flows through the measuring cell, it becomes partially oxidized or reduced. This process leads to an anodic or cathodic current, respectively, that is proportional to the concentration of the species being analyzed over a limited concentration range. This signal can, therefore, be converted into a chromatographic peak.

This type of detector has been used to analyze a wide range of both organic and inorganic ions in the ppb range. Using this technique, the analysis of environmental poisons—such as sulphide, cyanide and the phenols [10]—are particularly worth mentioning. A summary of some electroactive species and the working electrodes and potentials suitable for their analysis is given in Table 6-1.

TABLE 6-1: Electroactive species and the appropriate working electrodes and potentials that can be used for their analysis.

COMPOUND	WORKING ELECTRODE	WORKING POTENTIAL
HS^-, CN^-	Ag	0V
Br^-, I, SCN^-, SO_3^{2-}, $S_2O_3^{2-}$	Ag	0.2V
N_2H_4	Pt	0.5V
OCl^-	Pt	−0.2V

THE BASIC THEORY OF VOLTAMMETRY

The appropriate potential for amperometric detection of a given species can be obtained from voltammetric experiments. Voltammetry is the study, for analytical purposes, of current-voltage curves and the dependence of these curves on the nature of the electrode reactions and reactant concentrations. Of the many types of voltammetric measurements, those resulting from hydrodynamic and pulsed voltammetry are most easily applied to amperometry. For both these techniques, the substance being analyzed is dissolved in a solution of a base electrolyte that fulfills several conditions:

- By reducing the resistance of the solution, it ensures that the voltage drop, $i \cdot R_L$, is kept small. This condition is fulfilled when the base electrolyte concentration exceeds 0.1 mol/dm^3.

- It also ensures that the depolarizors (anions or cations) are not transferred to the electrodes. To fulfill this condition, the base electrolyte concentration should be some 50 to 100 times that of the depolarizors.

With only a few exceptions, it is the chlorides, chlorates, or perchlorates of the alkali or the alkaline earth metals, alkali hydroxides, or carbonates, or quaternary ammonium compounds that have been used as the base electrolytes.

Generally, the following reaction occurs during the amperometric detection of a species A:

$$A \rightleftharpoons B + ne^- \tag{122}$$

As the reaction proceeds, the species A donates n electrons to the working electrode and is transformed into a species B. The relationship between the concentrations of the oxidized and reduced species and the applied potential is given by the Nernst equation:

$$E = E_o + \frac{0.059}{n} \lg \frac{[B]}{[A]} \tag{123}$$

E: The working potential.

A and B: The equilibrium concentrations of the two species at the electrode surface.

E_o: The standard potential which exists when the concentrations of A and B are equal.

(Since the concentrations involved in chromatographic analyses are generally very small, the concentrations of the individual species can be used as a reasonable approximation in Equation 123, instead of their activities.)

In order to obtain an optimum current yield, the working potential must be chosen to ensure that A is completely oxidized to B. Thus, as can be seen from Equation 123, the working potential must always be greater than the standard potential E_o.

Recording a voltammogram of the oxidation of a substance A can be achieved using hydrodynamic voltammetry. For this technique, the substance, dissolved in a base electrolyte solution, is pumped through a cell containing an appropriate working electrode. The potential applied to the working electrode is continually increased and the current flowing is recorded.

Figure 6-4 shows the characteristic sigmoidal curve obtained from such an experiment. The form of the curve can be explained in terms of the concentration of the species being oxidized at the surface area of the electrode (see Figure 6-5).

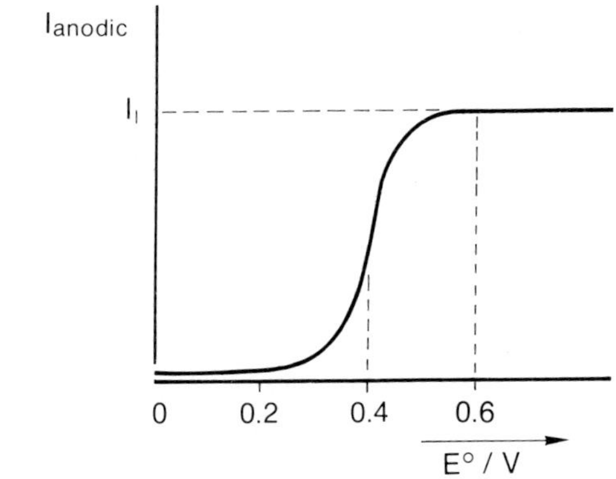

FIGURE 6-4.
Voltammogram recorded during the oxidation of a substance, A.

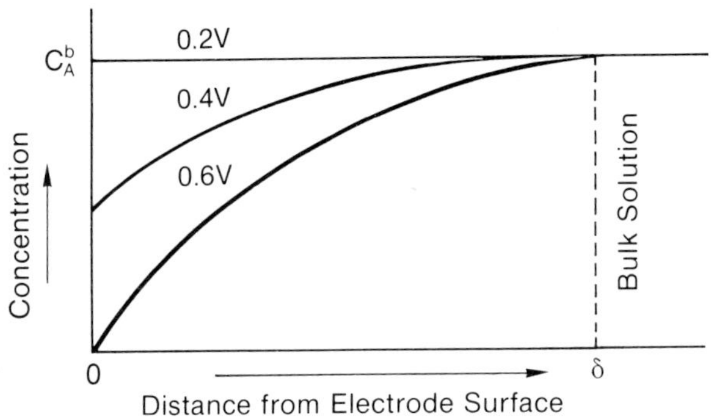

FIGURE 6-5.
The dependence of the concentration c on the distance between the surfaces of the electrodes at different applied potentials.

Up to potential E_A, only a small current (the rest or base current) flows. This arises from the charging of the electrical double layer at the electrode and from the reaction of impurities in the solution. The concentration of A at the electrode surface is equal to that in the bulk solution. Above potential E_A, the concentration of B increases with respect to that of A until the standard potential E_o is reached when the concentration of A and B, at the electrode surface, are equal. However, a few μm from the electrode surface, the concentration of A remains constant. The distance between the electrode surface and the point at which the concentration of A reaches its concentration in the bulk solution is known as the thickness of the diffusion layer δ. Due to the concentration gradient, the transport of A to the electrode surface within this layer is solely diffusion controlled. As the potential is raised towards E_B, the current increases and reaches a maximum at E_B. Thereafter, despite further increases in the potential, the current remains essentially constant since, as fast as A reaches the electrode it is oxidized to B. The amount of current is controlled by the rate of transport of A to the electrode surface because this is the slowest process. The current at this stage is called the limiting diffusion current and is proportional to the concentration of A in the bulk solution. Thus:

$$I = \frac{n \cdot F \cdot A \cdot D \cdot c}{\delta} \tag{124}$$

F: Faraday's constant.
A: The surface area of the electrode.
D: The diffusion coefficient of the substance A.

In a three electrode measuring cell, pulsed amperometry is carried out on a nonflowing solution containing both the base electrolyte and the species to be analyzed. The potential is applied in pulses and is increased between each pulse. After each pulse, the current is measured. The resulting voltammogram is the same as that shown in Figure 6-4. Immediately following a pulse in the diffusion controlled plateau region, the current flow is large due to the molecules of A at the electrode surface being oxidized immediately. The molecules of A further away from the electrode must first diffuse to its surface before they can be oxidized. This leads to a drop in the strength of the current between potential pulses as is shown in Figure 6-6.

Basically, there are two types of current developed during the oxidation of A to form B: the charging current and the Faradaic current. Whereas the charging current arises from the boundary layer between the electrode and the solution acting as a capacitor, the Faradaic current results directly from the electron transfer reaction. As can be seen from Figure 6-6, the charging current decreases exponentially with time. The relationship between the Faradaic current and time is given by an equation derived by Cotrell:

$$I(t) = \frac{n \cdot F \cdot A \cdot D^{1/2} \cdot c}{\pi^{1/2} \cdot t^{1/2}} \tag{125}$$

t: Time.

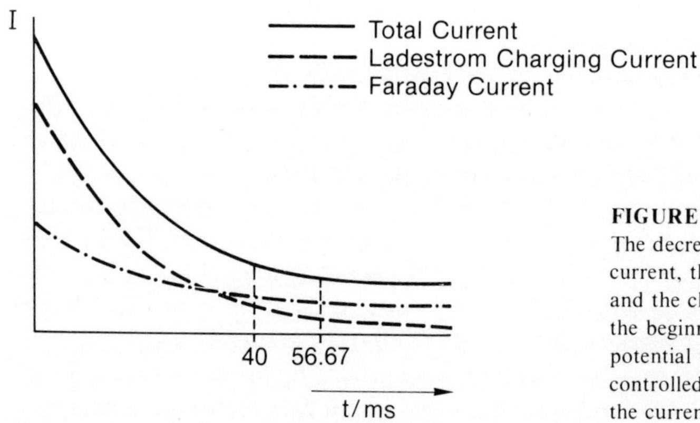

FIGURE 6-6. The decrease in the total current, the Faradaic current, and the charging current after the beginning of a pulse of a potential within the diffusion controlled, plateau region of the current-voltage curve.

Although the current strength is a function of $1/t^{1/2}$, it is directly proportional to the concentration of the ion being analyzed. The sum of both the charging and Faradaic currents is what is actually measured. Since the charging current is much greater than the Faradaic current immediately after a pulse, the current is usually not measured for at least 40 ms after the start of a pulse. Within this delay time, the ratio of the Faradic current to the charging current increases since the latter decreases much more rapidly.

AMPEROMETRY

Although the appropriate working potential for the proposed electrochemical reaction is obtained from voltammetric experiments, it is amperometry that is the actual detection method used in ion chromatography. One has to differentiate between amperometry carried out at a *single* potential and *pulsed* amperometry.

Amperometry at a Single Potential

This type of amperometry is the most frequently used electrochemical detection method for liquid chromatography. The method involves applying a constant d.c. potential across appropriate electrodes within a measuring cell. The theory associated with amperometry operating with a single potential is the same as that for hydrodynamic voltammetry, even though the potential remains constant.

The working potential is adjusted to the diffusion controlled, plateau region of the current-voltage curve (see above) for the ion being analyzed. If more than one ion is to be analyzed in a single run, the voltage must be adjusted so that the plateau region is reached for all the ions being analyzed. This type of amperometry is routinely used for the analysis of all the ions listed in Table 6-1.

Pulsed Amperometry

As a prerequisite, an amperometric detection of an electroactive species requires that the products of the oxidation or reduction will not precipitate on the electrode surface. Contaminated electrodes change their characteristics and this leads, in terms of amperometric detection for chromatography, to baseline drifting, greater background noise, and continually changing analytical sensitivity. This phenomenon is particularly prevalent during the amperometric detection of carbohydrates.

In contrast to amperometry with a single potential, pulsed amperometry uses three different working potentials — E_1, E_2, and E_3 — in a rapidly repeating sequence. The potentials can be applied for times t_1, t_2, and t_3 in 60 ms increments. Another difference to conventional amperometry is that, in the pulsed amperometric technique, the current flow is only measured for short periods of time. The use of an additional more positive and more negative potential allows the products of the electrode reactions to be removed from the electrode surfaces. The advantages of this technique are emphasized in Figure 6-7 showing the chromatograms obtained for a sugar alcohol mixture using both pulsed and continuous amperometry. The required working potentials and pulse times were derived from cyclic voltammetric experiments with the sequence shown in Figure 6-8.

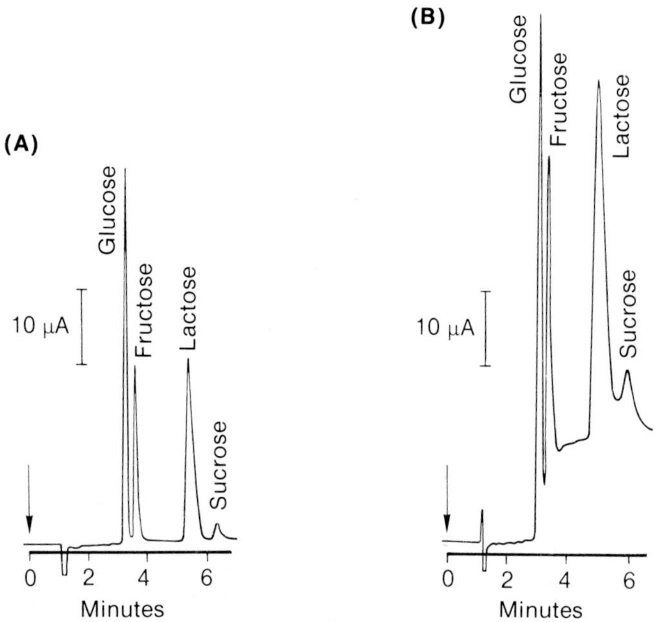

FIGURE 6-7.
The separation of sugar alcohols in a 1:100 dilution mixture of cocoa milk using a latex exchange column: (a) detection by pulsed amperometry with a Au working electrode (b) detection by conventional amperometry at a single working potential.

With the development of pulsed amperometric detectors (PAD) [11], a new measuring cell was designed that makes changing the working electrode considerably easier. This measuring cell has two halves joined by a sealing ring. In one half are the reference electrode and the capillary connections. The other half contains the working and counter electrodes. The latter can be exchanged as a single unit.

The choice of the appropriate working electrode for a given application is determined by the following factors:

- Potential limit of the working electrode in the eluant to be used.
- The possibility of the electrode being oxidized via the formation of complexes with components of the column effluent.

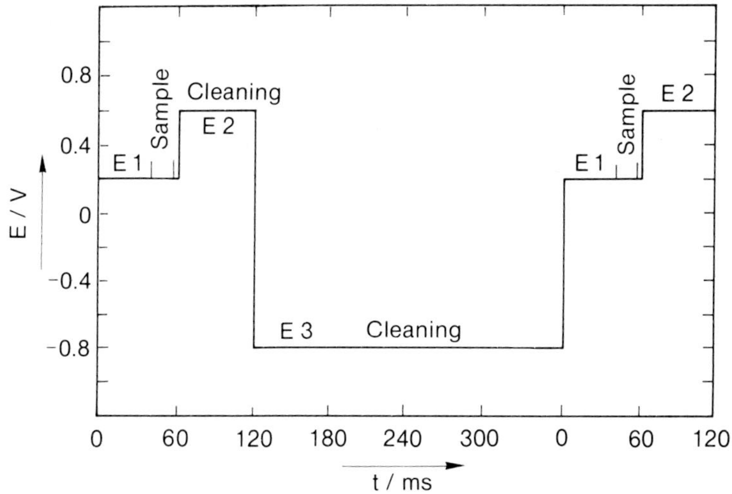

FIGURE 6-8.
Potential sequence for a Au electrode being used to detect carbohydrates.

- The kinetics of the electrochemical reaction.
- Background noise.

Potential Limit

The negative limit to the applied potential is given by that potential at which the base elecrtrolyte starts to be reduced. The positive limit to the applied potential is given by that potential at which the solvent, the base electrolyte, or the electrode start to be oxidized. In Table 6-2, the potential limits for the Au, Ag, Pt, and GC electrodes — in both acid and basic solutions — are summarized.

TABLE 6-2: The potential limits in acid and basic solution of some common working electrodes (reference electrode: Ag/AgCl).

WORKING ELECTRODE	SOLUTION (0.1 N)	POTENTIAL RANGE
Gold (Au)	KOH	−1.25V to +0.75V
	$HClO_4$	−0.35V to +1.10V
Silver (Ag)	KOH	−1.20V to +0.10V
	$HClO_4$	−0.55V to +0.40V
Platinum (Pt)	KOH	−0.90V to +0.65V
	$HClO_4$	−0.20V to +1.30V
Glassy Carbon (GC)	KOH	−1.50V to +0.60V
	$HClO_4$	−0.80V to +1.30V

The potential limits are especially dependent on the pH of the solution. In alkaline solution, more negative potentials can be applied than can be used in acid solution. On the other hand, more positive potentials can be used in acid than in alkaline solutions. For oxidation reactions at high positive potentials, GC and Pt electrodes are most suitable; whereas, reductions at strongly negative potentials can be carried out at GC, Au, and Ag electrodes.

Complex Formation

At positive potentials, there is a danger that complexes will form between silver or gold and some of the ions in the solution being analyzed. The formation of complexes is responsible for the lowering of the upper potential limit of these electrodes. However, complex formation can be advantageous when analyzing such anions as sulphide and cyanide.

Kinetics

If the reaction of A to B (Equation 122) is reversible, then the ratio of the concentrations of A and B at the surface of the electrode will attain an equilibrium position as defined by the Nernst equation (Equation 123). A prerequisite for reversibility is the preferential transfer of electrons between the electrode and the solute ion. For this reason, a compound is reduced or oxidized with different speeds, depending on the type of electrode used. If an electrode material which leads to a very slow electrochemical reaction is chosen, then the rate of reaction can only be varied by varying the applied working potential.

The difference between the equilibrium galvanic voltage and the corresponding electrode potentials is called the "overpotential." The phenomenon whereby the transport of material to and from the electrode is slower than the breakdown reaction leads to what is termed "concentration overpotential." The origin of a breakdown overpotential is a slow breakdown reaction; i.e., the transition of the particles from a particular form on one side to another form on the other side requires a definite activation energy. If the overpotential is greater than the potential limit for the electrode material being used, then an alternative material must be chosen.

Background Noise

Small irregularities in the flow rate, turbulence within the measuring cell, and small temperature fluctuations are common sources of background noise. Electrode materials that result in slower reactions lead to less background noise because their response times are longer than those of the electrodes at which the reactions taking place are more rapid.

6.4 FLUORESCENCE DETECTION

Detection via fluorescence is used, at the present time, only in connection with postcolumn derivatization for the analysis of amino acids and polyamines. The fluorescence is produced by exciting particular atoms with electromagnetic radiation so that they can then emit radiation as they relax. The wavelength of the emitted radiation is characteristic of the excited atom and the intensity is proportional to the concentration of the species in the solution.

Classical amino acid analysis involved the coloring of thin layer chromatograms with ninhydrin (1,2,3-indane trione) solution which led to a blue violet coloration of the spots containing amino acids. In 1973, Roth and Hampai [12] developed a fluorescence technique using o-phthalaldehyde and 2-mercaptoethanol that allowed the quantitative analysis of amino acids in the picomole range following their separation by ion exchange techniques. The fluorescence reagent is dissolved in a potassium borate buffer solution (pH = 10.4) and, with the aid of a PCR, mixed with the column effluent. The reaction, yielding an intensely blue, fluorescing complex, requires only a few seconds for completion. The wavelength used to excite the complexes is 340 nm and their emission can be measured at 455 nm. After oxidation with N-chloro-p-toluene sulphamide (Chloramin T), N-bromosuccinimide, or hypochlorite, it is also possible to analyze prolines (pyrrolidin-2-carboxylic acids) and hydroxyprolines (4-hydroxy-pyrrolidin-2-carboxylic acids). The excess reagent must, however, be destroyed before measuring the fluorescence since o-phthalaldehyde itself can be oxidized to give a fluorescing product.

LITERATURE

1. G. Ackermann and coworkers, "Elektrolytgleichgewichte und Elektrochemie," VEB Verlag für Grundstoffindustrie, Leipzig 1974.
2. M. Göbl, "Das Linearitätsproblem bei der Leitfähigkeits-Detektion in der Ionen-Chromatographie," GIT Fachz. Lab. **27,** 261–65 and 373–75 (1983).
3. R. P. Buck, S. Singhadeja, and L. B. Rogers, Anal. Chem. **26,** 1240–42 (1954).
4. R. N. Reeve, J. Chromatogr. **177,** 393–97 (1979).
5. U. Leuenberger, R. Gauch, K. Rieder, and E. Baumgärtner, J. Chromatogr. **202,** 461–68 (1980).
6. R. J. Williams, Anal. Chem. **55,** 851–54 (1983).
7. J. Stillian, "Trace Analysis via Post Column Chemistry in Ion Chromatography: Silica and ppb Calcium and Magnesium in Brines," Lecture Pittsburg Conference 1984.
8. H. Small and T. E. Miller, jr., Anal. Chem. **54,** 462 (1982).
9. R. A. Cochrane and D. E. Hillman, J. of Chromatogr. **241,** 392–94 (1982).
10. R. D. Rocklin and E. L. Johnson, "Determination of Cyanide and Sulfide, Iodide and Bromide by Ion Chromatography with Electrochemical Detection," Anal. Chem. **55,** 4–7 (1983).
11. P. Edwards and K. Haak, "New Pulsed Amperometric Detector for Ion Chromatography," Amer. Laboratory, April 1983.
12. M. Roth and H. Hampai, J. Chromatogr. **83,** 353 (1973).

7 Quantitative Analysis

7.1 GENERAL COMMENTS

The following discussion describes the various steps in developing a technique for the quantitative analysis of a particular material. It will be assumed that the chromatographic separation of the material has already been optimized.

The quantitative analysis of a material can be divided into the following individual working steps:

- Taking the sample.
- Sample preparation.
- Separation.
- Measurement.
- Signal processing.

The chemical analysis begins with taking the sample. The aim is to have a representative sample of the bulk material being analyzed. In particular, errors occur in the choice of a representative sample as well as during the storage and preparation of the sample. Special difficulties in obtaining a representative sample occur when the bulk material is an inhomogeneous solid, such as soils or soap powders. Physiological liquids, such as urine, are often collected over a longer period of time in order to ensure that the material being analyzed is, indeed, representative. Sample storage problems are particularly prevalent when the sample contains volatile, reactive, or oxygen sensitive components. Adsorption and desorption of sample material on the walls of the sample container must also be considered.

The sample preparation includes all operations concerned with converting the sample into an analysis form. Such operations include reducing the particle size of the sample (when necessary) as well as homogenization, digestion, dissolution, and filtration of the sample. Ideally, the sample solvent should be the same as the eluant used for its separation since this will reduce the negative, dead volume peak obtained when detection is made via conductivity. Before injection of the sample, the sample should always be filtered (0.25 or 0.45 μm) in order to avoid blocking of the analytical column by insoluble particles.

Injection of the sample usually takes place with the aid of a sample loop, allowing the sample volume to be easily reproduced and also allowing easy automation of the injection procedure.

The subsequent chromatographic separation also contains a number of possible error sources. Some of these include the coelution of different substances as well as the possible decomposition of sample components in the mobile phase or on the stationary phase.

The measurement procedure, quantifying some physicochemical property of the eluting sample components (conductivity, UV absorption, etc.) immediately follows the chromatographic separation. Such measurements must be unambiguous and correlate, in a known way, with the chemical and structural composition of the sample. The measurements recorded are then processed and summarized in an analytical report.

7.2 ANALYTICAL CHEMICAL INFORMATION PARAMETERS

When processing the results of the measurements, it is essential that the relevant quantities for determining the chemical composition of the sample are collected. These quantities, since they contain the essential analytical information, will be called the information parameters.

Chromatography involves the recording of a chromatogram from the continuous signal emitted by the detector. Qualitative information is contained in the retention time, t_{ms}, determined by the chromatographic process and dependent on the thermodynamic properties of the separation system and the solute. Quantitative information is contained in the area under the chromatographic peak. The relationship between the area and the quantity of material is a function of the detector and the measurement procedure. Of course, this presumes that the component being analyzed does, indeed, elute quantitatively from the analytical column being used.

The peak area and the retention time of a peak, i, correspond (according to Huber) to the zero moment, m_{0i}, and the first moment, m_{1i}, respectively, of the function describing the peak distribution.

$$m_{0i} = \int_{-\infty}^{+\infty} y_i(t)\,dt \tag{126}$$

$$m_{1i} = \int_{-\infty}^{+\infty} t\, y_i(t)\,dt \tag{127}$$

m_{0i}: The peak area A_i.
m_{1i}: The retention time.
t_{ms}: The component i.
y_i: The signal intensity of the component i.
t: Time.

The second moment, m_{2i}, of the function $y_i(t)$ for a peak, i, is given by the peak variance with respect to time σ_{ti}^2. Thus,

$$m_{2i} = \int_{-\infty}^{+\infty} t^2 \, y(t) \, dt \tag{128}$$

7.3 MEASURING THE PEAK AREA

The first step in the quantitative evaluation of a chromatogram is to measure the peak areas or to measure quantities proportional to the peak areas (peak height, peak area via triangulation, or peak height times its width at half maximum height). The various methods for determining the peak area differ in terms of their:

- Correctness.
- Precision.
- Applicability for asymmetric peaks.
- Time/cost required for the evaluation.

MANUAL EVALUATION OF PEAK AREA AND PEAK HEIGHT

Peak Height

As can be seen from Figure 7-1, the peak height is obtained by measuring the distance between the baseline and the peak maximum. Any baseline drift can be allowed by interpolating the baseline between the start and the end of the peak. The peak height is only porportional to the peak area if the peak form is constant. Thus, generally speaking, this is only an approximate method. However, if overlapping peaks are to be quantified, the peak height is a more reliable measure than the peak area since it tends to be independent of the resolution. The precision of the measurement of peak height as a measure of material concentration—depending on the quality of the calibration—is between 1 and 2% (relative).

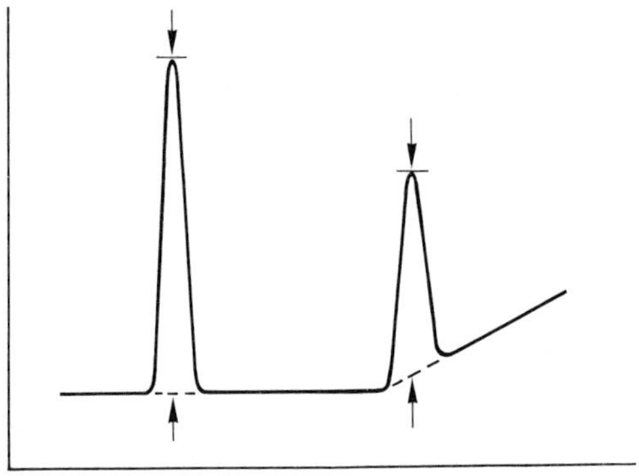

FIGURE 7-1.
Measurement of peak height.

Peak Area via Multiplication of Peak Height and Peak Width at Half Maximum Height

Chromatographic peaks can often be approximated by a triangle. Hence, the calculation of the area under the peak (see Figure 7-2), is given by:

$$A = H \cdot w_{1/2} \qquad (129)$$

H: The peak height.
$w_{1/2}$: The width of the peak at half the peak height.

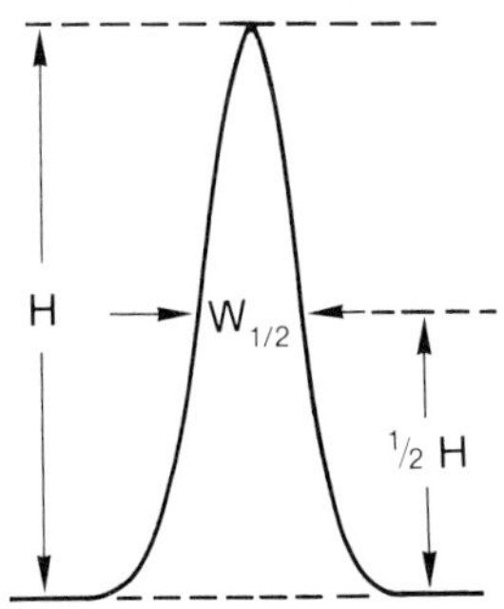

FIGURE 7-2.
Peak area calculated from peak height × peak width at half height.

Since the width of the peak at the baseline is influenced by adsorption and tailing effects, the width at half peak height is used for such calculations. Even so, this method should only be used for symmetrical peaks. The area calculated by this method is, indeed, proportional to the concentration of the species responsible for the peak, but tends to be somewhat smaller than the actual area. Furthermore, in order to measure accurately the peak width at half peak height, it is necessary to choose an appropriately fast recorder speed.

Triangulation

This technique is also based on the approximation that the chromatographic signal is effectively a triangle. In order to calculate the area, tangents are drawn to the peak above the points of inflection on the sides of the peak. The peak height is then measured between the intersection of the two tangents and the baseline, as shown in Figure 7-3. The peak width is given by the distance between the intersections of the two tangents with the baseline. The peak area is then calculated using

$$A = \tfrac{1}{2}\, B \cdot H \tag{130}$$

B: The peak width as given from the intersection of the two tangents with the baseline.

H: The peak height, measured from the intersection of the two tangents and the baseline.

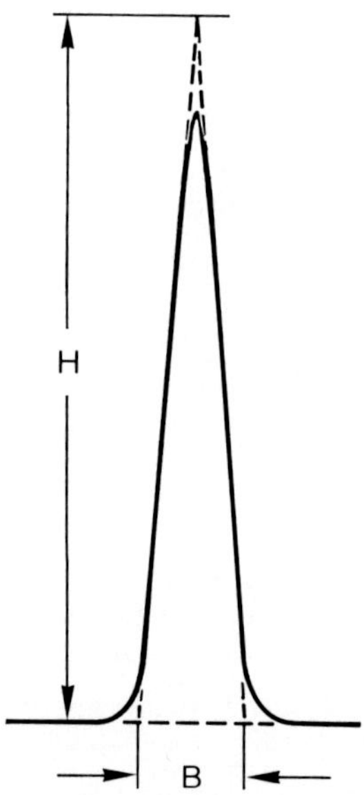

FIGURE 7-3.
Triangulation.

This triangulation procedure has the same limits as have been discussed above. In addition, it has the disadvantage that tangents need to be drawn and that small errors in drawing the appropriate tangents can lead to greater errors in the value of the peak height. Thus, this method is not recommended.

ELECTRONIC DETERMINATION OF THE PEAK AREA

Digital Integrators

With the aid of a digital integrator, the peak area and retention times can be measured and the baseline drift can be compensated automatically. In a digital integrator, a voltage to frequency converter (V/F converter) produces an outgoing signal whose frequency is proportional to the incoming signal. During an integration, the impulses from the V/F converter are counted—the total number being proportional to the peak area. In this way, digital integrators can measure the areas of peaks produced by signals varying from several microvolts to about 1 volt. Thus, the linear range covers some five orders of magnitude. Additionally, the recorder signal can be attenuated independently from the integrator signal. Digital integrators have the advantage of greater precision and the automatic transformation of chromatographic signals into a numerical form. Nevertheless, it is important to ensure that the parameters used are correctly chosen so that the integrator does not make incorrect adjustments for baseline drift or jumps that can lead to errors in peak area calculations.

A summary of the methods available for determinig peak areas and the quantities proportional to the peak areas is given in Table 7-1.

TABLE 7-1: A summary of the various methods of assessing peak areas.

METHOD	TIME REQUIRED	MAXIMUM PRECISION OF ROUTINE ANALYSIS	CORRECTNESS	APPLICABILITY FOR ASYMMETRICAL PEAKS
Peak height measurement	Short	0.5% (dependent on peak symmetry)	Good (for symmetrical peak shape)	Not suited
Peak height × width at half maximum height	Short	2%	Good (for constant and approximate symmetrical peak shape)	Less suited
Triangulation	Short	3%	Good (see above)	Less suited
Electronic integration	—	0.1%	Very good (dependent on the algorithm for curve fitting, peak detection and baseline drift)	Suited

Computer

Although digital integrators can be used for the quantification of peak areas and retention times, their use for controlling the operation of a chromatograph and for more advanced data processing is limited. For this reason, interest in developing a combination of chromatograph and dedicated microcomputer has grown. The result is the newest 2300i series of ion chromatographs. The microcomputer carries out the following jobs:

- Control and supervision of the complete chromatographic system.
- The calculation of peak areas and retention times.
- Calculation of the sample composition on the basis of available calibration and other data already stored.

The chromatographic conditions are typed into the microprocessor that then adjusts the starting conditions and starts the program. The necessary information for the evaluation (calibration and integration parameters) of the chromatogram is simply typed into the microprocessor. Provided all the necessary information is available, the microprocessor yields a full analytical report at the end of the chromatographic run.

7.4 DETERMINING THE COMPOSITION OF THE SAMPLE

In order to derive the composition of the sample from a chromatogram, the peak area, A_i, is presumed to be proportional to the amount of substance, Q_i, which gives rise to the peak. Thus,

$$A_i = f(Q_i) = \bar{S}_i \cdot Q_i \tag{131}$$

$S_i = (dS/dQ)_{iQ}$: The response factor of the detection and recording systems for the component, i, when the amount present is equal to Q.

$\bar{S}_i$: The average response factor obtained from the slope of the regression line whose equation is $A = A_o + \bar{S}_i \cdot Q$.

The value of $\bar{S}_i$ must be found by calibration. However, the average sensitivity, $\bar{S}_i$, is not, generally, constant; but rather varies with the amount of substance, Q_i. Thus, it is important not to rely on a single calibration with an arbitrarily chosen quantity. Either the amount used for calibration, Q_{i0}, must be chosen so that it is close to the amount in the unknown sample, Q_{ix}, or a calibration function must be derived by measuring a number of calibration samples, $Q_{i1} \ldots Q_{in}$, whereby the range is chosen to include the amount likely to be found in the unknown sample. The calibration function describes the relationship between the peak area and the amount of substance in the sample and is not, generally, a linear function. Nevertheless, the calibration function can often be approximated by the linear function given in Equation 131. Only if larger concentration ranges are included does the deviation from linearity become noticeable.

With the aid of modern, electronic integrators, the calibration can be carried out using four different methods:

- Area normalization.
- Internal standard.
- External standard.
- The method of standard additions.

AREA NORMALIZATION

The method of area normalization assumes that all components of the sample are eluted from the analytical column and that they are all detected. If the response factors are not allowed for, this method can only be used for calibration of sample components with equal response factors.

The method of area normalization is most widely used in the gas chromatography of hydrocarbons. It has only limited applicability to ion chromatography, since response factors for different components are rarely equal.

The calculation of the area of the unknown peak (X in Figure 7-4) is carried out as follows:

$$\%X = \frac{A_x \cdot 100}{A_x + A_y + A_z} = \frac{A_x \cdot 100}{\sum_{i=1}^{n} \cdot A_i} \tag{132}$$

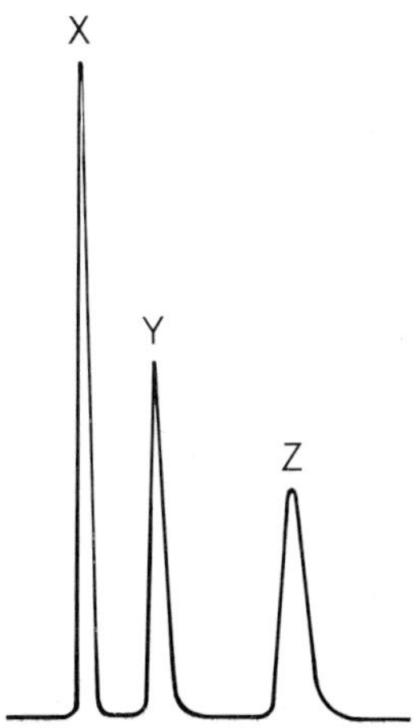

FIGURE 7-4.
Normalization of peak areas.

INTERNAL STANDARD

For this method, a given concentration of standard substance is added to the sample being analyzed. The composition of the unknown is then calculated by comparison with the peak area of the standard. With this method, it is not essential that all the components of the unknown are either eluted or detected. Thus, this method is suitable for samples containing some components that need not be analyzed. For the internal standard, the following criteria are important:

- The standard must be chemically stable under the chromatographic conditions to be used.
- The substance should elute close to the unknown.
- The peak of the standard must be baseline separated from its neighboring peaks.
- The response factor of the substance must be known or able to be determined.
- The standard should have similar structure, response factor, and concentration as the unknown.
- It must be very pure.

For calibrations using an internal standard, it is essential that the compounds to be analyzed are available in extremely pure form for use as references. In order to determine the response factors, a solution—containing both the internal standard and the substances to be analyzed—is made up and chromatographed. The response factors can then be calculated from:

$$f_x = \frac{A_x \cdot W_{is}}{A_{is} \cdot W_x} \qquad (133)$$

f_x: The response factor for the component x.
A_x: The peak area.
W_x: The concentration.
A_{is}: The peak area of the internal standard.
W_{is}: The concentration.

A calibration curve (A_x/A_{is} vs. W_x) is constructed from results obtained from several solutions containing the same concentration of internal standard and differing concentrations of the substances to be analyzed.

Having obtained a calibration curve, the appropriate concentration of the internal standard is added to the unknown sample and the sample chromatographed. The composition of the sample can then be calculated using:

$$\%X = \frac{A_x \cdot f_x \cdot W_{is} \cdot 100}{A_{is} \cdot W_s} \qquad (134)$$

Employing peak areas and providing care when calibrating, the relative accuracy of this method can be as good as ±0.1%.

EXTERNAL STANDARD

Direct comparison of the area of a peak in an unknown sample with that produced from a known concentration of the same substance is the method most frequently used in ion chromatography. The method requires that the injection volumes are accurately reproducible and that the chromatographic conditions can be held constant. Errors due to variation in the volumes injected can be neglected if a sample loop is used. The method demands that all the substances to be analyzed are, indeed, available as references. In practice, various standard solutions with concentrations similar to that of the unknown are made up and chromatographed. A plot of the resulting peak areas against concentration of the standard (see Figure 7-5), allows the calculation of the response factor for the substance being investigated.

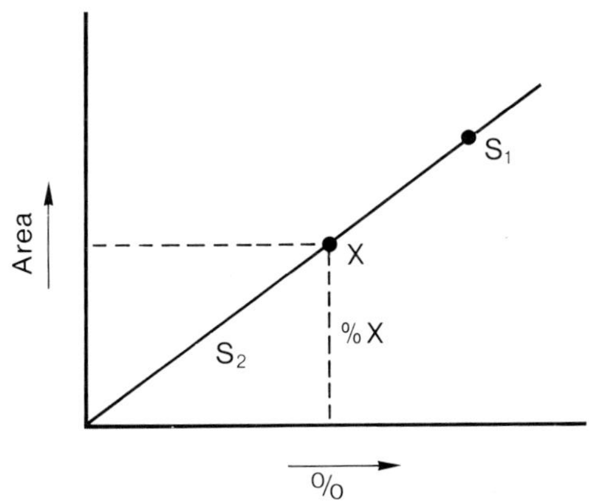

FIGURE 7-5. Calibration using an external standard.

The area of the peak of the unknown should lie between two of those measured for known concentrations, so that the unknown concentration can be determined by interpolation. Due to the limited linearity of these calibration curves, extrapolation is not recommended. The slope of the calibration curve shown in Figure 7-5 is equal to the response factor for the particular substance. It is both substance and detector specific. Assuming that the injection volume is accurately reproducible, the relative accuracy of this method — under routine conditions — is ±0.3%.

STANDARD ADDITIONS

The method of standard addition involves a combination of internal *and* external standard methods. It is especially useful when carrying out ion chromatography of samples with complex matrices. After chromatographing the unknown sample, a known amount of a particular component is added and the sample is rechromatographed. When calculating the concentration of the particular substance, the change in volume must be taken into account. As an internal standard for correcting errors due to sample volume, any of the other peaks in the chromatogram can be used. Evaluation is made by using Equation 135:

$$Q_i = \frac{\Delta Q_i}{\dfrac{(A + \Delta A)_{i2}}{A_{i1}} \cdot \dfrac{A_{s1}}{A_{s2}} - 1} \qquad (135)$$

Q_i: The concentration of the substance of interest before addition of the standard.

ΔQ_i: The added concentration.

$A_{i1}, (A + \Delta A)_{i2}$: The peak areas of component i in the chromatograms 1 and 2.

A_{s1}, A_{s2}: The peak areas of the reference peak in the two chromatograms.

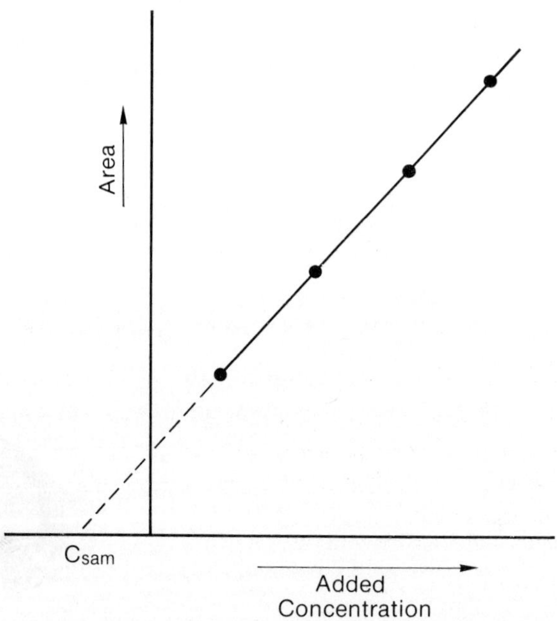

FIGURE 7-6. Calibration using the method of standard addition.

Evaluation of the concentration of the substance being analyzed can also be made graphically. As can be seen from Figure 7-6, one plots the peak area of the substance being analyzed as a function of the amount added. Assuming that the detector (conductivity, UV absorption, etc.) response is proportional to the concentration of the species responsible for the peak, such a plot gives a straight line which can be extrapolated to its negative intercept of the x-axis to give the concentration in the sample.

7.5 STATISTICAL TREATMENT OF QUANTITATIVE ANALYTICAL DATA

It is of paramount importance that analytical results be evaluated in terms of their *accuracy* and *precision;* i.e., in terms of their systematic and statistical errors.

The *accuracy* of an analytical result is a measure of the difference ε_{ai} between the measured value and the true value. It is only possible to determine the *accuracy* of a result if the true value is known. The term *precision* of an analytical result describes the differences between individual measurements made on the same sample under identical chromatographic conditions. The precision is usually characterized by the standard deviation σ_{Ai} of the average value μ_{Ai}. If the measurement is repeated n times, it is possible to calculate estimated values for the average, $\hat{\mu}_{Ai}$, and standard deviation, s_{Ai}, by using:

$$\hat{\mu}_{Ai} = \frac{\sum_{i=1}^{n} A_i}{n} \tag{136}$$

and

$$S_{Ai} = \frac{\sum_{i=1}^{n} (A_i - \mu_{Ai})^2}{n-1} \tag{137}$$

As the number of measurements increases, the estimated values $\hat{\mu}_{Ai}$ and s_{Ai} approach the values of μ_{Ai} and σ_{Ai}, respectively.

The standard deviation is a measure for the size of the random error associated with the analysis. The random error depends both on the method and the composition of the sample. The relative, random error ε_{pi} is a measure of the precision of the method analysis. As ε_{pi} decreases, the precision increases.

$$\varepsilon_{pi} = \frac{\sigma_{Ai}}{\mu_{Ai}} \tag{138}$$

With the aid of the relative, random error, the analytical chemist can quickly judge whether the results—independent of their absolute values—are within the precision limits of the method.

The statistical reliability of a result can also be given in terms of the *confidence interval T* and its *confidence limits.* The interval indicates that the degree of probability of correctness, P%, of all individual measurements averaging $\hat{\mu}_{Ai}$, can be expected to lie between $\hat{\mu}_{Ai} + T$ and $\hat{\mu}_{Ai} - T$. The value of T is a measure of the quality of the crude data without having to know the individual values. The interval T can be calculated from:

$$T = s \cdot t \tag{139}$$

T: The student's factor (dependent on the degree of probability, P%, and the degrees of freedom, f = n − 1).

In order to calculate T, a particular value of P% has to be chosen and this has to be considered when judging the result. This value must also be recorded:

$$\text{Result} = \hat{\mu}_{Ai} \pm T \text{ (units)}; (\pm s_{Ai}; P\%; n)$$

In contrast to the confidence limit, T, of the individual values, the average value is defined as $T/\sqrt{n}$.

If a number, m, of average values, μ_i, is determined (each of which involves n individual measurements), the estimated value of the standard deviation, $s_{\mu i}$, of the average is given by:

$$s_{\mu i} = \frac{s_{Ai}}{\sqrt{m}} \tag{140}$$

This implies that the estimate of the standard deviation of the average value decreases relative to that of the individual measurements as m increases.

NALIMOV'S OUTLIER TEST

If an individual measurement is very different from other measurements on the same sample, it is possible (using Nalimov's test) to determine whether this result is an element of a homogeneous set of data.

This involves calculating a value, r*, by inserting the suspicious data point in the equation:

$$r^* = \frac{|x^* - \hat{\mu}|}{s} \cdot \sqrt{\frac{n}{n-1}} \tag{141}$$

A decision can be made by comparing the value r* with values of r_i given in standard tables [1]. It should be noted that the value of r_i is dependent on the number n of measurements made and the statistical reliability given by P%.

Example: From an analysis of the concentration of chloride in a sample, the following data was obtained:

$$x_1 = 30.4 \text{ ppm}$$
$$x_2 = 30.0 \text{ ppm}$$
$$x_3 = 30.5 \text{ ppm}$$
$$x_4 = 30.2 \text{ ppm}$$
$$x^* = 29.1 \text{ ppm}$$

$$\text{Mean value: } \mu = 30.04 \text{ ppm.}$$

$$\text{Standard deviation: } s = \pm 0.313.$$

For Nalimov's test, the degree of freedom, f, must be calculated using:

$$f = n - 2 \tag{142}$$

with n = 5, f = 3. Inserting this and the value of the suspect analysis into Equation 141 gives:

$$r^* = \frac{|\,29.1 - 30.04\,|}{0.313} \cdot \sqrt{\frac{5}{5-1}}$$

$$= 3.357$$

From the r-tables, the values of r_i can be found. Thus, for f = 3, the values for r_i are:

r (95%): 1.757
r (99%): 1.918
r (99.5%): 1.982

Comparison of r* with the values of r_1 allow a decision to be made as to whether x* is an outlier or not:

$r(95\%) < r^* < r(99\%)$: Probable outlier
$r(99.9\%) < r^* < r(99.9\%)$: Significant outlier
$r^* < r(99.9\%)$: Highly significant outlier

If, after the comparison described above, the suspected outlier is shown to be a highly significant outlier, the average and standard deviation should be recalculated without including the value of the outlier:

Mean value: $\mu = 30.28$ ppm

Standard deviation: $s = \pm 0.049$

Clearly, the corrected values are very different from those calculated including the outlier.

REGRESSION ANALYSIS

As already described in Section 7.4, the response factor describes the relationship between the peak area and the concentration of the substance responsible for the peak. Since systematic deviation from a linear relationship is not uncommon, it is important to have a measure of the linearity of this function. For this, it is possible to use the maximum difference between the linear function and the true function within a particular concentration range. The linear range is then defined by a maximum signal strength below which the relative deviation is less than a given value.

The equation of the straight line, $Y = a + b \cdot x$, can be obtained by a linear regression analysis of the data. The coefficients a and b are calculated according to Gauss's method. This gives the straight line that leads to the smallest sum of the squares of the differences between the individual points and the straight line; i.e., for n values, x_i, y_i:

$$\sum_{i=1}^{n} (y_i - Y_i)^2 = \sum_{i=1}^{n} [y_i - (a + b \cdot x_i)^2] \qquad (143)$$

has a minimum value.

Thus,

$$b = \frac{\sum_{i=1}^{n} (x_i - \bar{x}) \cdot (y_i - \bar{y})}{\sum_{i=1}^{n} (x_i - \bar{x})^2} = \frac{\sum_{i=1}^{n} x_i y_i - \frac{1}{n}\left[\sum_{i=1}^{n} x_i \cdot \sum_{i=1}^{n} y_i\right]}{\sum_{i=1}^{n} x_i^2 - \frac{1}{n}\left[\sum_{i=1}^{n} x_i\right]^2} \qquad (144)$$

and

$$a = \overline{y} - b \cdot \overline{x} \tag{145}$$

In these equations, $\overline{x}$ and $\overline{y}$ are the average values of x_i and y_i. A measure of the actual linear correlation between x_i and y_i is given by the correlation coefficient $r_{x,y}$ that can be calculated from:

$$r_{x,y} = \frac{\sum_{i=1}^{n} (x_i - \overline{x}) \cdot (y_i - \overline{y})^2}{\sqrt{\sum_{i=1}^{n} (x_i - \overline{x})^2 \cdot \sum_{i=1}^{n} (y_i - \overline{y})^2}}$$

$$= \frac{\sum_{i=1}^{n} x_i y_i - \frac{1}{n} \sum_{i=1}^{n} x_i \cdot \sum_{i=1}^{n} y_i}{\sqrt{\sum_{i=1}^{n} x_i^2 - \frac{1}{n}\left(\sum_{i=1}^{n} x_i\right)^2} \sqrt{\sum_{i=1}^{n} y_i^2 - \frac{1}{n}\left(\sum_{i=1}^{n} y_i\right)^2}} \tag{146}$$

This coefficient is independent of the units of the individual data points and can assume any value from −1 to +1. If $r_{x,y} = 1$, the correlation is exactly linear.

CALCULATION OF THE CONFIDENCE INTERVAL

In order to calculate the confidence interval for the coefficient b, it is first necessary to choose the level of confidence required γ (95%, 99%). The value for c in the equation:

$$F(c) = \frac{1}{2} \cdot (1 + \gamma) \tag{147}$$

is then found from the t-tables with n−2 degrees of freedom. Then one calculates:

$$l = c \cdot \left(\frac{\sqrt{a}}{s}\right) \cdot \sqrt{(n-1) \cdot (n-2)} \tag{148}$$

The confidence interval is then:

$$b - l \leq b \leq b + l \tag{149}$$

The concept of confidence interval is further demonstrated by the plot shown in Figure 7-7. The relative error L in the slope b is given by

$$L = \frac{1}{b} \cdot 100 \qquad (150)$$

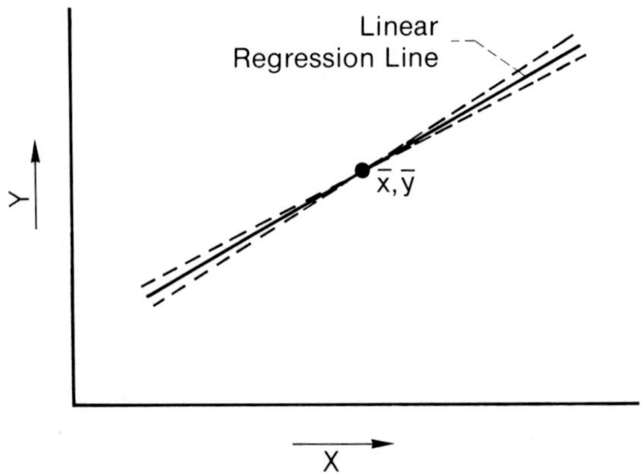

FIGURE 7-7.
Plot showing the relevance of the confidence interval.

LITERATURE

1. R. Kaiser and G. Gottschalk, "Elementare Tests zur Beurteilung von Meβdaten," Bibliographisches Institut Mannheim, 1972.

ADDITIONAL READING

W. J. Dixon and F. J. Massey, "Introduction to Statistical Analysis," McGraw-Hill, New York 1957.

8 Instrumentation

In the summer of 1981, a new series of Dionex ion chromatographs—the Series 2000i—was introduced. These instruments are liquid chromatographs specially designed for the separation and quantitative analysis of a wide variety of ionic species.

As already explained, ion chromatography is often carried out with highly corrosive eluants such as HCl, H_2SO_4, etc. Thus, straightforward HPLC systems with capillaries and pumps constructed from stainless steel are not suitable for ion chromatography since, with time, even rust-free stainless steel is attacked by such elements. This leads to a number of contamination problems:

- Metal ions have high affinities for the stationary phases of ion exchange columns. A contamination by such ions leads to a significant loss of separation capability [1].
- When using amperometry, contamination due to metal ions leads to increased background noise.
- The analysis of transition and heavy metals at trace levels is made very difficult if the eluant is contaminated with metal ions.

For these reasons, chromatographs were developed in which the parts contacting the mobile phase are constructed from inert materials such as Tefzel, Teflon, polypropylene, Kel-F, sapphire, and ruby. The only exception is the eluant valves in the analytical pump module that contain a small cone made from shockhardened 716 steel. This item (which has a very small surface area and is situated on the low pressure side of the pumps) has, up to now, not been observed to suffer excessive corrosion.

The modular concept of the Series 2000i ion chromatograph permits the basic model 2010i to be upgraded to a two channel, automatic system. The latter permits two separations to be carried out simultaneously. The main points of the individual modules are discussed below.

8.1 THE ANALYTICAL PUMP

Since a constant flow rate is essential for reproducible chromatographic analyses, a double piston pump was constructed. Because it is electronically controlled, the pump guarantees maximum precision and pulse free operation. Flow rates between 0.1 and 9.9 cm^3/min in steps of 0.1 cm^3/min can be selected. The stability of the flow rate is ca. 0.25%. Since the analytical pumps can be used to produce pressures up to 130 bar, they can also be used for simple HPLC applications using chemically bound reversed phase columns. An alarm switch ensures that the pressure in the system remains within predetermined maximum and minimum pressures. Deviations outside these limits lead to the pump turning itself off. With the aid of switches, it is possible to choose any of six different eluants. If deionized water is selected, the four eluant valves and the eluant manifold are rinsed to reduce contamination. The control of the eluant valves is electronic and can, therefore, be automated. A priming valve is situated alongside and coupled, via a one way valve, to the pump heads. This valve arrangement has several functions:

- If the valve into the pressure transducer is opened, the pump heads can be primed by forcing the eluant through the pump with the aid of a plastic syringe.
- After changing to a new eluant, it is possible, with the aid of a syringe, to draw out any air that has collected in the eluant tubing.

Both pump heads are equipped with conventional one way check valves. The pressure transducer is constructed from stainless steel and the surfaces covered with a polymer film. The eluant is carried from this module to the pumps via plastic capillary tubing with an internal diameter of 0.4 mm.

A particular feature of the analytical pump is the electronic drive unit that, via feedback loops from each individual piston, ensures constant flow rate and backpressure. From the block diagram shown in Figure 8-1, it can be seen that the pump is driven by a direct current motor and the associated drive unit is connected to an electronic unit. The electronic unit can analyze error messages and, when necessary, regulate the pump's operation. The electronic drive unit controls the feedback loops from the two pump heads. One of these feedback loops compares the actual pressure with the preset value and adjusts the speed of the pump accordingly. This simple control loop leads to an excellent and constant flow rate. The second feedback loop compares the actual and preset backpressures. The preset value is calculated by multiplying the backpressure that is measured over a given length of time by a factor derived from the difference between the actual and the desired flow rates. If this factor is equal to one, then the actual backpressure is equal to what it should be: the flow rate is the same as that desired. If this factor is not equal to one, then the motor receives an error message and its speed is regulated accordingly. When the pump is initially turned on, this second feedback loop is not activated immediately. Only after the backpressure has become reasonably stable does it come into operation and the system is then in its "ready" mode. This state can be recognized by the control lamp beginning to glow. The extremely pulse free flow required by both UV and amperometric detectors is guaranteed by these pumping systems.

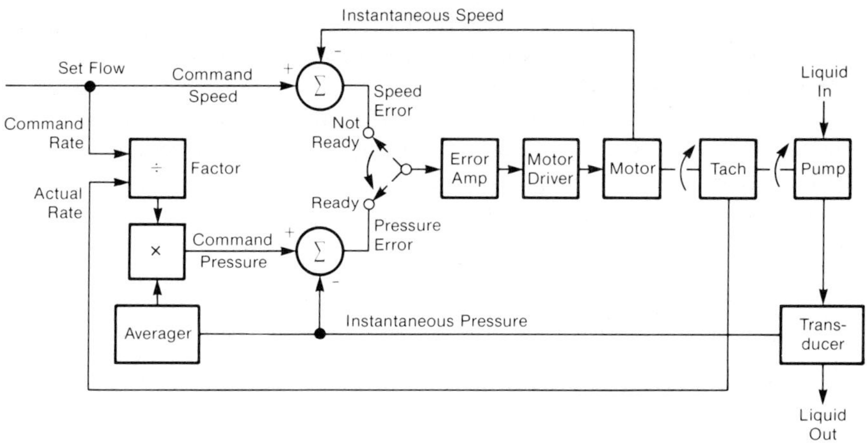

FIGURE 8-1.
Block diagram showing the principles of the electronic control of an analytical pump.

8.2 THE CHROMATOGRAPHY MODULE

Those parts of the chromatography module that come into contact with the mobile phase are also constructed from nonmetallic materials to avoid corrosion problems. The chromatography module is, even in its simplest form, laid out for a simultaneous, two column operation and can be purchased with either three or six switching valves. This unit has sufficient room for several complete column systems consisting of precolumn, separation column, and suppressor column. The column cabinet has a plexiglass door, so that the columns are easily accessible and exchanging columns presents no problems. In the upper part of the cabinet, it is possible to include a column heater which is necessary for the analysis of carbohydrates. The temperature of this heater can be adjusted between room temperature and 99° C in increments of 1° C. The back of the column cabinet can also be opened to allow the operator access to the various valves and measuring cells attached to it. This arrangement allows the measuring cells to be isolated from the temperature variations in the laboratory. Smaller variations in temperature can be electronically compensated (see Section 8.3).

The close proximity of the detection cells to the columns also allows the dead volume of the system to be kept to a minimum by avoiding unnecessary lengths of capillary (0.3 mm ID) between the units. A one column system has three valves after the columns; situated vertically one above the other. These valves, also constructed from nonmetallic materials, can cope with internal pressures up to 130 bar and are operated pneumatically. The valves are controlled electronically so that both manual operation via switches and automatic operation are possible. The top valve is the injection valve and the ancillary valves "A" and "B" can be used for automatic regeneration of suppressor columns or for various other column switching functions (e.g., during coupled HPICE/HPIC). On the front of the chromatography module, there are two injection parts and an additional four parts (per channel) for eluant, regenerating medium, and various detectors. The two channels can be operated either in series or independently.

8.3 THE CONDUCTIVITY DETECTOR [2]

Determining the capability of a conductivity detector involves assessing the following criteria:

- Linearity.
- Resolution.
- Noise.
- Temperature compensation.

These four criteria will be discussed using the detector developed for use in conjunction with the Series 2000i chromatographs as an example.

LINEARITY

In traditional conductivity detectors, a sinusoidal alternating voltage is applied across the electrodes. The formation of electrical double layers at the electrodes leads to a capacitative resistance developing in addition to the simple ohmic resistance. The former leads to a phase shift of the applied sinusoidal voltage. With analog detectors, the simple ohmic resistance cannot, therefore, be measured and nonlinearity results when changing measurement ranges. The introduction of a bipolar pulse technique solves this problem. It can be seen from the current-time profile of a single pulse (shown in Figure 8-2), that the initial signal is deformed by the charging and dissipation of capacitance. The measurement of the ohmic resistance is made digitally when the capacitance is fully charged. The negative portion of the pulse, during which electrode polarities are reversed, serves simply to discharge the "ionic cloud" that has built up at the electrode. This measurement technique is free of nonlinearity between measurement ranges.

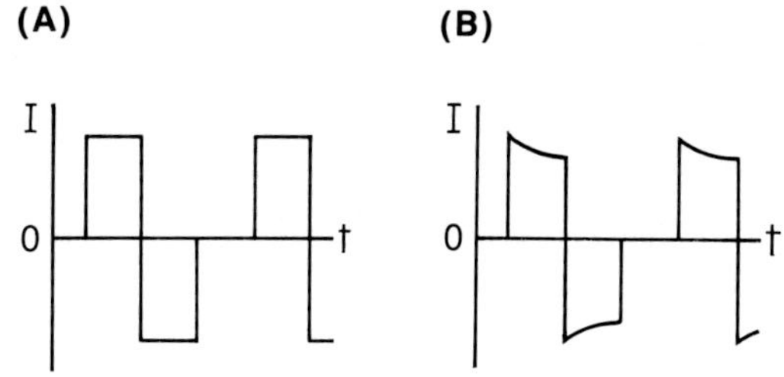

FIGURE 8-2.
Current versus time profile for a bipolar pulse: (a) input signal (b) output signal.

RESOLUTION

In comparison to the straightforward conductivity bridges commercially available, detectors coupled with a microprocessor have a considerably improved sensitivity. In order to be able to process the signals by a microprocessor, it is first necessary to digitize the signal. This is accomplished by an analog to digital converter (A/D converter) that usually contains a 16-bit processor. This implies that signals between 0 and 2^{16} (65,536) can be processed. Thus, the total span of measurement, usually between 0 and 10^4 μS/cm, can be divided into 65,536 divisions. This results in only 6 divisions being available for the range 0 to 1 μS/cm and the resulting chromatograms are not suitable for quantitative evaluation. A considerable improvement in the sensitivity is achieved by limiting the 16-bit processor to the range from 0 to 5 μS and, at the same time, using an 11-bit A/D converter to produce an offset signal which can be subtracted from the actual measurement (see Figure 8-3). If the conductivity becomes greater than 5 μS, the offset signal increases so that the resulting signal remains less than 5 μS. With this technology, the sensitivity of the detector can be increased by a factor of 2^{11} (= 2,048). The total resolution is thus $1 : 134 \times 10^6$; i.e., for the range 0 to 1 μS, 13,400 divisions are available.

It is only since the advent of microprocessor technology that conductivity detection of ions has been able to penetrate the middle and lower ppb ranges without the need to preconcentrate the sample.

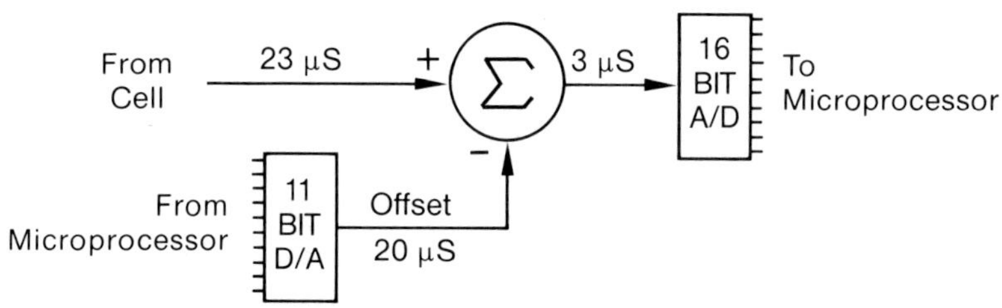

FIGURE 8-3.
Circuit diagram enabling the resolution of a conductivity detector to be improved.

BACKGROUND NOISE

Following the improvement in the sensitivity of conductivity detectors, considerable efforts were made to minimize the background noise. In this respect, it is worth considering the input signal for the A/D converter shown in Figure 8-4 which has a significant level of noise superimposed on the desired signal. In order to reduce the noise, traditional A/D converters measure the signal at different times (1 point measurements) and produce an output with less noise (see Figure 8-4a).

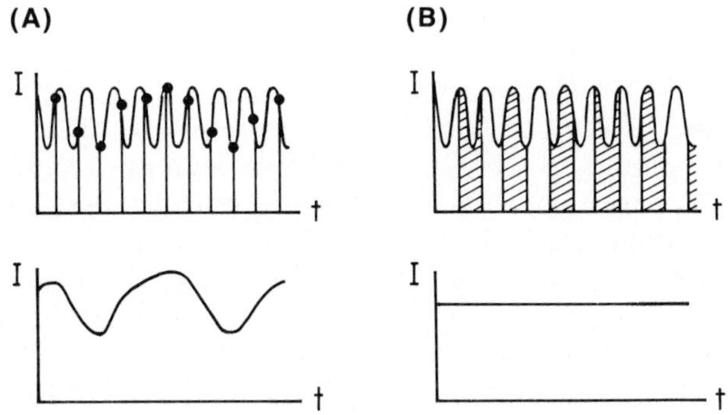

FIGURE 8-4.
Comparison of the noise suppression by: (a) single point, direct measurement and (b) integrating measurement. $\Delta t = 200$ ms.

A considerably better method of suppressing unwanted background noise is employed by the so-called "integrating" A/D converters. A converter of this type measures the input signal over a particular time interval, Δt, and calculates the area under the signal. The area is then divided by the time, Δt, to obtain an average signal. The time interval, Δt, used is 200 ms; so that between measurements, the processor can carry out the necessary, complicated calculations. Noise arising from the primary power source is also filtered by this type of processor, since the time interval corresponds to a frequency of 6 Hz which is an integral factor of the main frequency (60 Hz).

A further reduction in the background noise is achieved via the process of digitizing the baseline offset. In all analog detectors, an offset voltage is produced to allow for baseline compensation. This voltage is, however, not constant and tends to contain a level of noise. In the newer conductivity detectors, this offset voltage is digitized by the processor for storage and continually subtracted from the measured voltage. The stored value is constant and invariable until a new value is inserted. It is, thus, free of noise.

COMPENSATION OF TEMPERATURE VARIATIONS

Since the conductivity of an electrolyte solution is a function of temperature, it is important that measures are taken to correct for minor fluctuations in the temperature of the measurement cell. With the introduction of microprocessor technology, it became practical to do away with complex thermostating and to compensate for the temperature variations mathematically. The basis of such a mathematical correction of the measured conductivity of an infinitely dilute electrolyte solution at a temperature corresponding to a value of 25° C is given by Equation 151:

$$\lambda^{\circ}_{\vartheta} = \lambda^{\circ}_{25}\, e^{[k(\vartheta - 25)]} \qquad (151)$$

$\lambda^{\circ}_{\vartheta}$ and λ°_{25}: The equivalent conductivities of the ions being measured at temperatures ϑ and 25° C, respectively, in an infinitely dilute solution.

The k in Equation 151 is termed the temperature coefficient of the conductivity and can be determined experimentally. It is specific for a particular ion in a particular eluant. In aqueous solutions, the value of the experimental term can be approximated by 1.017. This means that the equivalent conductivity of an ion increases by some 1.7% for each °C rise in temperature.

The compensation of temperature variations by using Equation 151 assumes that the exact temperature of the solution in the measurement cell is known at all times. To achieve this, the cell is so constructed that it can include a small thermistor (0.5 mm diameter) hanging in the eluant flow. If the response time for such a thermistor is ca. 1.5 sec, the relative reproducibility of the temperature measurements is ca. 1%. In this way, temperature changes in the eluant flow can be registered both quickly and accurately.

With these two quantities — the temperature coefficient of the conductivity and the actual temperature of the measurement cell — the microprocessor can continuously calculate the conductivity according to Equation 151 so that the values recorded are those corresponding to 25° C. It should, however, be noted that, during analyses in the sub ppb range, the absolute conductivity can only be determined using a multichannel analyzer that can take account of the individual temperature coefficients for each of the individual ions being analyzed. In the range between ppb and ppm, the errors arising from using incorrect temperature coefficients is negligibly small compared with those arising from the injection volume.

8.4 THE CONTROL MODULE

In conjunction with an automatic sampler, the "AutoIon 100 Controller" completely automates either a single or a two channel setup. The basis of the AutoIon 100 is two 8748 microprocessors, four relays, and two a.c. outputs per channel.

The front of the control unit is analogous to the individual modules comprising the system so that each module has its own, individual section with the appropriate keys for controlling that module. With an additional switch, the individual channels can be chosen and independently operated so that, during an analysis on one channel, the other channel can be programmed.

Nine different programs per channel can be stored and a total of 420 program steps are available. After installing the programs in the "program" mode, the sequencing of the steps to be carried out and repeated with respect to the samples to be analyzed is decided while the unit is in the "schedule" mode. Each program can be repeated up to 99 times; but, if this is not enough, an endless loop can be programmed using the code-word "or" (override). Repeating a complete program sequence is also possible by using the code "J" (jump).

Changing to the "'run" mode readies the control unit to carry out the operations programmed as soon as the "start" key is pressed. The operations can be stopped at any time using the "stop" key and continued simply by depressing the "start" key. Restarting the program sequence is affected simply by depressing the "reset" key.

LITERATURE

1. J. Dutrieu and Y. A. Delmotte, Techscan, Sept. 1981, p 8 (DuPont de Nemours).
2. M. Göbl, GIT **27**, 261–65 and 373–75 (1983).

9 Applications

The determination of ionic species in solution is a classical analytical problem, solved by employing a wide variety of experimental methods. Whereas the analysis of cations has, for some time, been possible using a range of rapid and sensitive techniques (AAS, AES, ICP, etc.), a highly sensitive method for analyzing anions was lacking. Conventional wet chemical techniques — such as titration, photometry, gravimetry, turbidimetry, and colorimetry — for the analysis of ions — such as chloride, phosphate, nitrate, and sulphate — are labor intensive, time consuming, and are, sometimes, susceptible to interference.

The major areas of application are:

- Water analysis.
- Semiconductor industry.
- Electroplating industry.
- Medicine.
- Agriculture.
- Power generation.
- Pharmaceutical industry.
- Paper manufacture.
- Mining and metal purification.
- Food and beverage industry.
- Environmental protection.

In Tables 9-1 and 9-2, the ion chromatographic separation and detection systems are summarized, together with some of the ions that can be analyzed by the individual systems.

Particular importance has been achieved by ion chromatography as a method for analyzing all types of water samples [1]. Thus, the ions chloride, phosphate, nitrate, and sulphate (which determine the quality of water), can be simultaneously analyzed in less than ten minutes. Figure 9-1 shows a typical analysis of drinking water from Weiterstadt, West Germany. Great sensitivity (detection limit using *direct injection*: 10 ppb), as well as the possibility of automation, were the major reasons for the rapid acceptance of the method. Using ion chromatographic methods, it is also possible to analyze more complex water samples — such as wastewater or soil extracts — without any problems. The result of an

TABLE 9-1: A summary of ion chromatographic separation systems.

SEPARATION SYSTEM	MECHANISM	FUNCTIONALITY OF THE RESIN	RECOMMENDED ELUANT	SPECIES TO BE ANALYZED
HPIC Anions	Ion exchange	$-\overset{+}{N}R_3$	Na_2CO_3/ $NaHCO_3$	F^-, Cl^-, Br^-, I^-, SCN^-, CN^-, PO_2^{3-}, PO_3^{3-}, PO_4^{3-}, $P_2O_7^{4-}$, $P_3O_{10}^{5-}$, NO_2^-, NO_3^-, S^{2-}, SO_3^{2-}, SO_4^{2-}, $S_2O_3^{2-}$, $S_2O_4^{2-}$, $S_2O_6^{2-}$, $S_2O_8^{2-}$, OCl^-, ClO_2^-, ClO_3^-, ClO_4^-, SeO_3^{2-}, SeO_4^{2-}, AsO_4^{3-}, CrO_4^{2-}, MoO_4^-, WO_4^-, carbohydrates, etc.
Cations	Ion exchange	$-SO_3^-$	HCl; m-Phenylene-diamine · (2 HCl/HCl)	Li^+, Na^+, NH_4^+, K^+, Rb^+, Cs^+, Mg^{2+}, Ca^{2+}, Sr^{2+}, Ba^{2+}, small aliphatic amines.
		$-SO_3^-/-\overset{+}{N}R_3$	Oxalate/ citrate tartrate	Fe^{2+}, Fe^{3+}, Cu^{2+}, Ni^{2+}, Zn^{2+}, Co^{2+}, Pb^{2+}, Mn^{2+}, Cd^{2+}, Al^{3+}.
HPICE	Ion exclusion	$-SO_3^-$	HCl Octane sulphonic acid	Aliphatic carboxylic acids, carbonate, borate, citrate, malonate, succinate, lactate, alcohols.
MPIC Anions	Ion pair formation and adsorption	Neutral	NH_4OH, TMAOH, TPAOH, TBAOH	In addition to those anions listed under HPIC: metal cyano complexes, BF_4^- and anion surfactants.
MPIC Cations	Ion Pair formation and adsorption	Neutral	HCl, hexane-sulphone acid, octane sulphone acid	In addition to those cations listed under HPIC: alkyl amines, alkanol amines, quaternary ammonium compounds, choline, cation surfactants.

TABLE 9-2: A summary of ion chromatographic detection systems.

DETECTION	PRINCIPLE	APPLICATION
Conductivity	Electrical conductivity	Anions and cations with pK_a or pK_b <7.
Amperometry	Oxidation and reduction at a Ag/Pt working electrode	Anions and cations with pK_a or pK_b >7 (S^{2-}, CN^-, OCl^-, etc.).
Pulsed amperometry	Oxidation and reduction at a Ag/Pt/Au and glassy carbon electrode	Alcohols, carbohydrates, phenols, nitro aromatics, etc.
UV/Vis detection in connection with postcolumn reaction	UV/Vis light absorption	UV active anions and cations, heavy metals after reaction with PAR and Tiron, polyvalent anions after reaction with iron-nitrate, silicate after reaction with molybdate, Ca^{2+}/Mg^{2+} in NaCl after reaction with Arseneazo I.
Fluorescence in connection with postcolumn reaction	Excitation and emission	Amino acids and polyamines after reaction with o-phthalaldehyde.

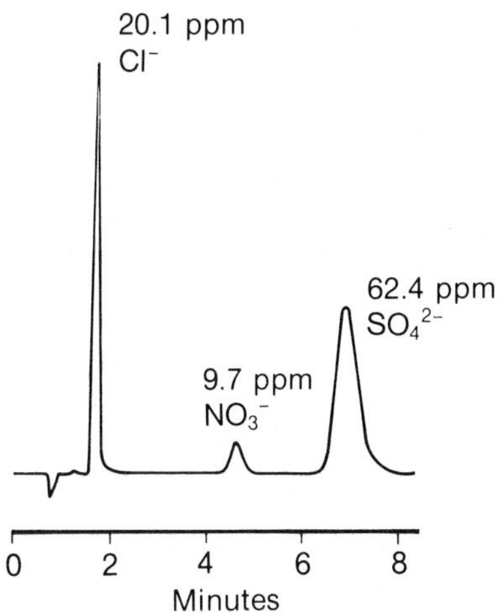

FIGURE 9-1.
An analysis of the anions in the drinking water from Weiterstadt. **Column:** HPIC-AS3. **Eluant:** 0.0028 mol/dm^3 $NaHCO_3$ + 0.0022 mol/dm^3 Na_2CO_3. **Flow rate:** 2.3 cm^3/min. **Detection:** Suppressed conductivity. **Sample:** 50 μl, dilute 1:2.

analysis of the wastewater from an isopropanol synthesis is shown in Figure 9-2 and demonstrates the separation of monoisopropyl sulphate and inorganic sulphate. The ion selective electrodes that can also be used for wastewater analysis are characterized by short working lives and a lack of selectivity. Recent publications from Darimont [1], Schwedt [2], and Merriweather [3], demonstrate that the results obtained using ion chromatographic techniques are comparable with those obtained by conventional techniques.

In addition to the analysis of water samples, ion chromatography has become important in five other areas of application. These are discussed in the rest of this chapter.

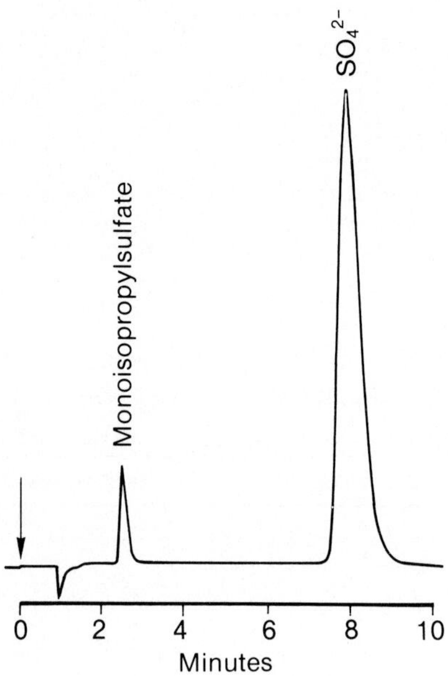

FIGURE 9-2.
An analysis of the anions in the wastewater from an isopropanol synthesis. **Column:** HPIC-AS3. **Eluant:** 0.0028 mol/dm^3 $NaHCO_3$ + 0.0022 mol/dm^3 Na_2CO_3. **Flow rate:** 2.3 cm^3/min. **Detection:** Suppressed conductivity. **Sample:** 50 μl, diluted 1:5000.

9.1 ION CHROMATOGRAPHY APPLICATIONS IN POWER GENERATION

The evaluation of water, steam, and distillate quality is the most important application of ion chromatography within power laboratories. Up to now, conductivity was the most widely used monitoring method. Routine plant control by ion chromatography allows leaks in condensers to be identified and the functioning of water treatment plants to be checked. Conductivity monitors are, however, nonspecific, since the size of the signal gives no information as to the nature of the impurity. Furthermore, sodium chloride or sulphate impurities (resulting from ingresses of cooling water or due to defective regeneration of the condensate cleaning filter), are not reliably detectable in the sub ppb range.

Using ion chromatography, the purest water can be analyzed for chloride/sulphate or its sodium content after concentration on a concentrating column. This can be achieved by attaching the instrument to a sampling loop so that a well defined amount can be pumped onto the concentrating column. Such an arrangement is substituted for the usual 50 μl sample loop attached to the injection port as shown in Figure 9-3. A second concentrating column, attached between connections 4 and 8, serves to equalize the pressures as the injection valve is switched. After the injection valve is switched, the concentrated sample from the concentrating column is forced onto the separating column [4]. Tandem switching with two concentrating columns can allow simultaneous analysis and loading.

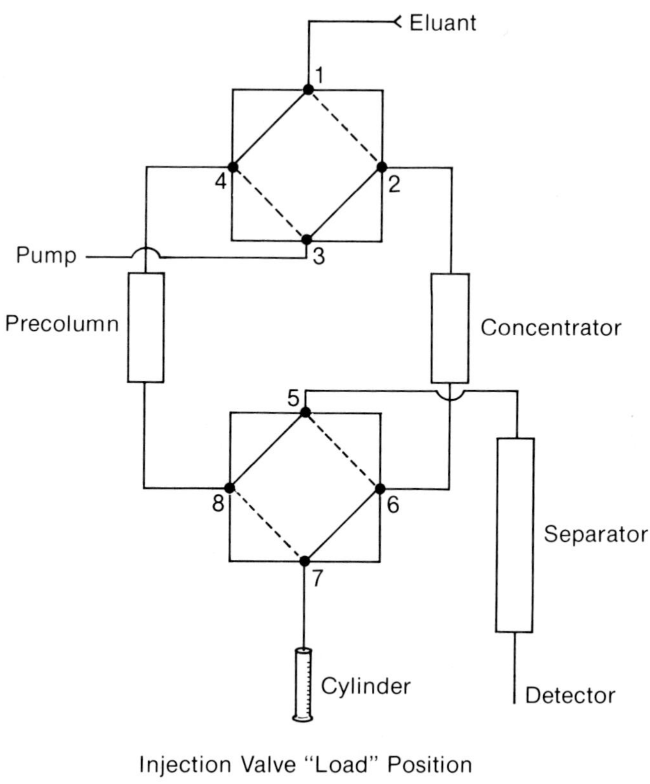

FIGURE 9-3.
Schematic showing the valves arrangement for a concentration technique.

With the valve arrangement shown in Figure 9-4, both direct injection onto a sample loop and a concentration technique can be carried out simply by switching the appropriate valve. Such a switching system is particularly useful when both pure water samples and more concentrated solutions are regularly analyzed on the same system; for example, when cooling water and deionized water are to be analyzed routinely. Figure 9-5 shows a chromatogram obtained from an analysis of "pure water" containing ca. 0.2 ppb chloride and 2 ppb sulphate after a sample of 40 cm^3 had been concentrated. With this technique, the detection limits for chloride, sulphate, and sodium could be improved from the level of 10 ppb—obtainable using *direct injection*—to ca. 0.1 ppb—after *concentration* [5]. The chromatogram from a sample of cooling water is shown in Figure 9-6.

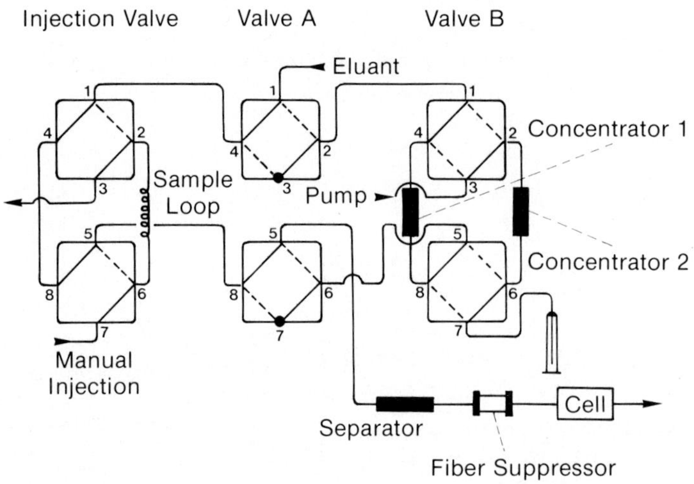

FIGURE 9-4.
Schematic showing the valves arrangement for a system to be used for both concentration and direct injection techniques.

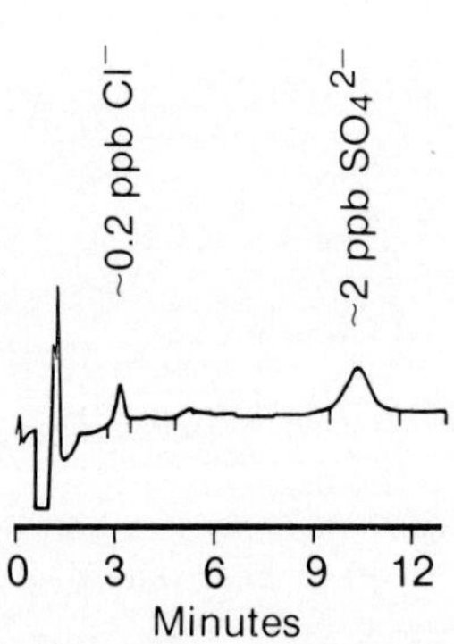

FIGURE 9-5.
An analysis of the anions in "pure" water after concentration. **Column:** HPIC-AS2 with an AG2 concentrator column. **Eluant:** 0.003 mol/dm^3 Na_2CO_3 + 0.002 mol/dm^3 NaOH. **Flow rate:** 2.3 cm^3/min. **Detection:** Suppressed conductivity. **Sample:** 40 cm^3 "pure" water.

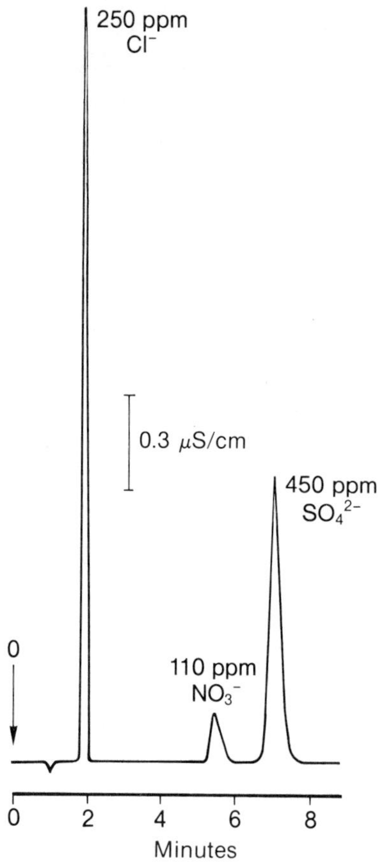

FIGURE 9-6.
An analysis of the anions in cooling water. **Column:** HPIC-AS4. **Eluant:** 0.0028 mol/dm^3 $NaHCO_3$ + 0.0022 mol/dm^3 Na_2CO_3. **Flow rate:** 2 cm^3/min. **Detection:** Suppressed conductivity. **Sample:** 10 μl.

To avoid overloading the high performance column, the cooling water is diluted 1:10 with deionized water. Additionally, direct injection of the cooling water is not worthwhile due to the nonlinearity of the conductivity detector's response at high concentrations.

Another important application of ion chromatography is as a control for the washwater circulation in flue gas desulphuration plants. The basic principle is that the sulphur containing species in the flue gas are absorbed by the aqueous phase. The washwater contains a number of inorganic acids which can be analyzed using ion chromatography [6]. During flue gas desulphuration using lime, sulphite is oxidized to sulphate. As can be seen in Figure 9-7, sulphate can be analyzed simultaneously with the other mineral acids. Recombination of two hydrogen sulphite radicals leads to the formation of dithionate, $S_2O_6^{2-}$, that can also be analyzed ion chromatographically [7]. The affinity of this ion for the stationary phase of latex anion exchange columns is so strong that only strong eluants—such as 0.05 mol/dm^3 sodium perchlorate—are effective. Because of

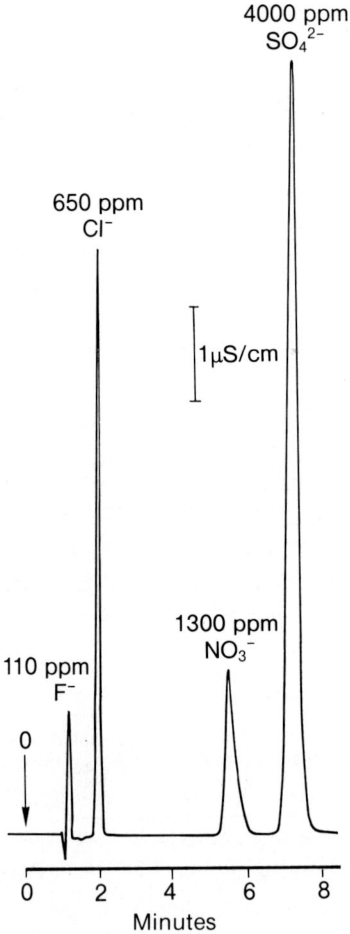

FIGURE 9-7.
The inorganic mineral acids in the washwater from a smoke desulphuration plant. **Column:** HPIC-AS4. **Eluant:** 0.0028 mol/dm^3 $NaHCO_3$ + 0.0022 mol/dm^3 Na_2CO_3. **Flow rate:** 2 cm^3/min. **Detection:** Suppressed conductivity. **Sample:** 50 μl, diluted 1:100.

the high ionic strength of the eluant, sensitive conductivity detection becomes impossible. As an alternative, UV detection at 240 nm can be used. Somewhat more elegant is the analysis of such ions with ion pair chromatography using tetrabutyl ammonium hydroxide as the ion pair reagent. The retention times of the ions are determined by the concentration of acetonitrile (the organic modifier) in the eluant. Figure 9-8 shows a representative chromatogram from an analysis of dithionate in the washwater from a flue gas desulphuration plant.

The formic acid added to some washing solutions to adjust the pH can be separated from other anions and analyzed in its formate form in samples of dissolved gypsum.

Ion chromatography can also be used to control the removal of the nitrogren oxides from flue gases. The most important parameter—the ratio of the con-

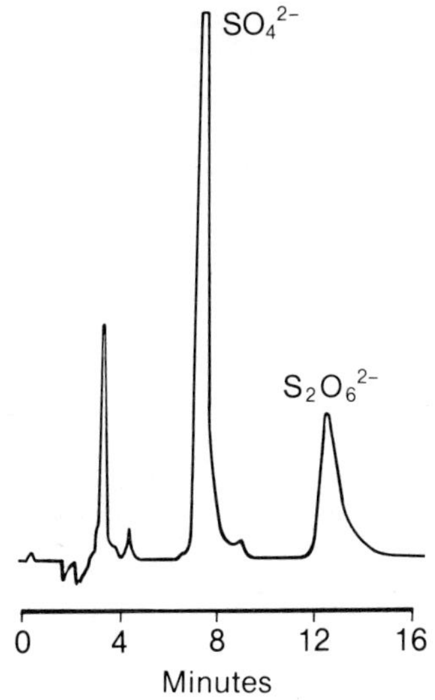

FIGURE 9-8.
An analysis of the dithionate in the washwater from a smoke desulphuration plant. **Column:** MPIC-NS1. **Eluant:** 0.002 mol/dm^3 tetrabutyl ammonium hydroxide + 0.001 mol/dm^3 Na_2CO_3 + 15 vol.-% acetonitrile. **Flow rate:** 1 cm^3/min. **Detection:** Suppressed conductivity. **Sample:** 50 μl, diluted 1:100.

centrations of nitrite and nitrate—can be determined (see Figure 9-9) in less than 5 minutes.

The analysis of inorganic sulphur containing species is also of importance to coal analysis (coal liquification and coking), since the fossil fuels contain not only organic sulphur compounds (mercaptans, sulphides, polysulphides, etc.). For the analysis of di- and polysulphuric acids (not previously analyzable by ion chromatography), special ion pair chromatographic techniques have been

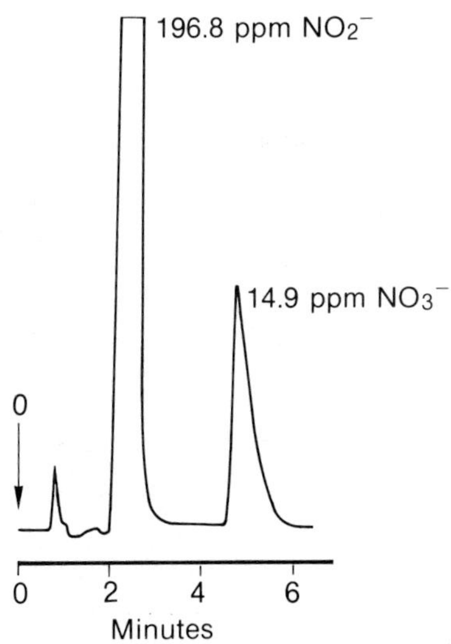

FIGURE 9-9.
An analysis of nitrite in run-off water. **Column:** HPIC-AS3. **Eluant:** 0.0028 mol/dm^3 $NaHCO_3$ + 0.0022 mol/dm^3 Na_2CO_3. **Flow rate:** 2.3 cm^3/min. **Detection:** Suppressed conductivity. **Sample:** 50 μl.

developed [8]. Under the chromatographic conditions listed under Figure 9-8, it is possible to analyze dithionite, $S_2O_4^{2-}$, which elutes before dithionate. In contrast, disulphite, $S_2O_5^{2-}$, decomposes in aqueous solution according to Equation 152:

$$S_2O_5^{2-} + H_2O \rightleftharpoons 2\ HSO_3^- \tag{152}$$

Disulphate, $S_2O_7^{2-}$, is also unstable and decomposes in an aqueous solution according to:

$$2\ S_2O_7^{2-} \rightleftharpoons 2\ SO_4^{2-} + 2\ SO_3 \tag{153}$$

In order to analyze the polythionates $S_nO_6^{2-}$, ($n > 2$) only the concentration of acetonitrile in the eluant needs to be increased. The only commercially available member of this class of compounds is tetrathionate. A chromatogram from a solution of this material is shown as an example in Figure 9-10. A separation of thiosulphate and dithionate is simultaneously possible. By similar techniques, dithionite was shown to be present as an impurity in tetrathionate. Peroxodisulphate, $S_2O_8^{2-}$, has a gross retention time (t_{ms} = 14.9 min) under these chromatographic conditions.

A new and interesting application for ion chromatography is the trace analysis of cooling water conditioners. These products are generally mixtures of polycarboxylic and polyphosphonic acids added to the cooling water as corrosion inhibitors. Using these conditioners, calcium hydrogen carbonate is stabilized in the regions of high thermal strain and lime deposits on metal surfaces are avoided.

Polyphosphonic acids can be separated on special latex anion exchange columns and detected, after complexation with iron (III), photometrically (see Chapter 3, Section 3.2, "Polyvalent Ions"). With this technique, a qualitative analysis of the inorganic and organic phosphates contained in the conditioner is possible. Figure 9-11 shows an example chromatogram of a commercially available product. As can be seen from this chromatogram, the product is a mixture containing two major components and several impurities. Since the concentration of conditioner added to the feed and circulating water is usually in the ppb range, the sensitivity of the *direct injection* technique is not adequate and the concentration technique described above must be used [9]. Using an initial volume of 2 cm^3, a chromatogram—such as that shown in Figure 9-12 from a sample of cooling tower water—can be obtained: easily detecting the presence of polyphosphonic acids. The other two signals in the chromatogram are due to chloride and sulphate. Both these ions form corresponding complexes with iron (III) and are, thus, also detected, especially since they are present in considerably higher concentrations.

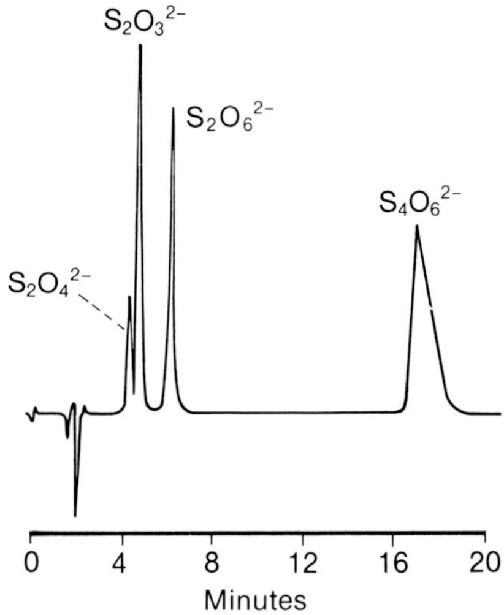

FIGURE 9-10.
An analysis of tetrathionate. **Column:** MPIC-NS1. **Eluant:** 0.002 mol/dm^3 tetrabutyl ammonium hydroxide + 0.001 mol/dm^3 Na_2CO_3 + 20 vol.-% acetonitrile. **Flow rate:** 1 cm^3/min. **Detection:** Suppressed conductivity. **Sample:** 50 μl, containing 20 ppm thiosulphate, 50 ppm dithionate, 100 ppm tetrathionate.

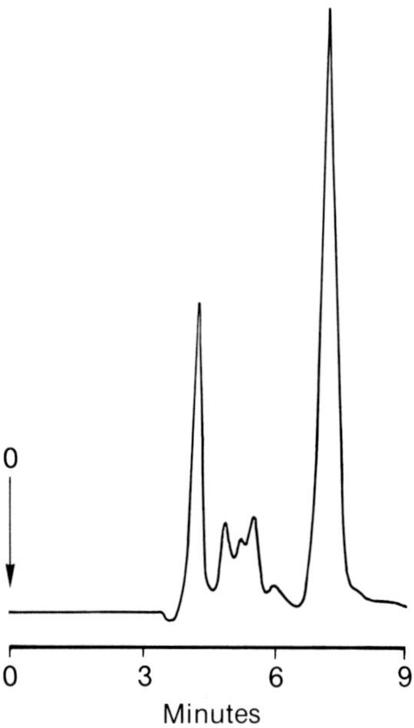

FIGURE 9-11.
An analysis of cooling water conditioner. **Column:** HPIC-AS7. **Eluant:** 0.03 mol/dm^3 HNO_3. **Flow rate:** 0.5 cm^3/min. **Detection:** Postcolumn reaction with iron (III) nitrate and UV detection at 330 nm. **Sample:** 50 μl, diluted 1:1000. **Sensitivity:** 0.05 aufs.

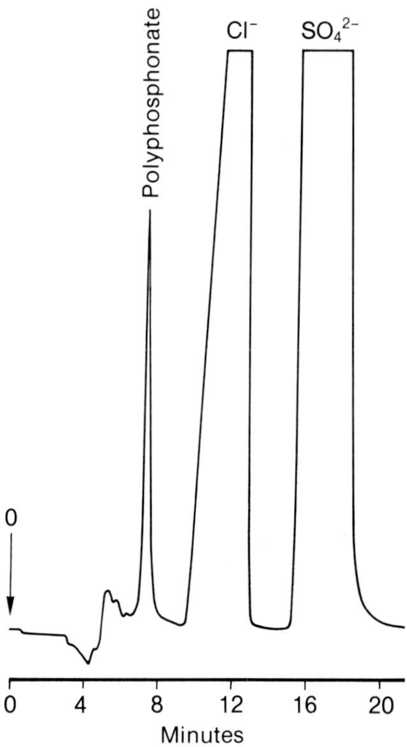

FIGURE 9-12.
An analysis of the conditioner in the circulating water from a cooling tower. **Column:** HPIC-AS7. **Eluant:** 0.03 mol/dm^3 HNO_3. **Flow rate:** 0.5 cm^3/min. **Detection:** Postcolumn reaction with iron (III) nitrate and UV detection at 330 nm. **Sample prior to concentration:** 2 cm^3.

A quantitative analysis is, at the moment, extremely difficult because of a lack of exact information about the composition of these conditioners. After identification of the major components by comparison with commercially available polyphosphonates, it should be possible (using ion chromatography) to quantitatively analyze the conditioners in cooling water so that more exact dosing will be possible.

9.2 ION CHROMATOGRAPHY IN THE ELECTROPLATING INDUSTRY

Within a very short time, the electroplating industry has become one of the most important areas of application for ion chromatography. The routine control of the plating solutions, and improvements in the quality of the plating as a function of the chemical composition, are paramount among the various analyses made in this sector. With increasing production costs for the relatively high value products, it became necessary to apply novel analytical methods to improve the level of understanding of the relatively complex chemical processes taking place in the plating baths.

Ion chromatographic analytical techniques allow rapid and reliable analysis of the major components—as well as reaction and decomposition products—in the plating baths. Compared to the wet chemical methods previously used which were, in some cases, nonspecific, the advantage of ion chromatography lies in the selectivity of the stationary phase and the detection systems employed. Thus, in almost all cases, sample preparation is reduced to simple dilution with deionized water and filtration. A larger number of applications [10, 11, 12, 13] in the electroplating field have been carried out for plating baths for both electroless and electrolyte deposition. They are summarized in Table 9-3.

Two examples given in Table 9-3 demonstrate the typical cost savings made by employing modern analytical techniques. They are discussed in detail below.

ELECTROLESS NICKEL BATHS

In comparison to electrolytic processes, electroless deposition is becoming increasingly preferred by the electroplating industry for two principle reasons:

- Problems due to zones of higher and lower current density do not occur. Hence, more uniform films of higher quality can be prepared.
- Nonconducting materials can be directly plated.

Electroless plating requires the addition of a reducing agent and a catalyst to the plating bath. The reducing agent acts as electron donor for the reduction of the

TABLE 9-3: Areas of application and analytical examples in the galvanic industry.

TYPE OF PLATING BATH	EXAMPLES OF ANALYSES
Copper electroless	Determination of: Formic acid, tartaric acid, tri-ethanol-amine, EDTA and Cu $(EDTA)^{2-}$.
Copper sulphate electrolytic	Determination of: Cl^-/SO_4^{2-}-ratio, metallic impurities such as Ni^{2+} and Zn^{2+}, organic additives.
Copper pyrophosphate electrolytic	Determination of: Ammonium, orthophosphate and nitrate.
Copper cyanide electrolytic	Determination of: Cu^{2+} after preparation with strong acids, $Fe(CN)_6^{3-}$, $Fe(CN)_6^{4-}$.
Nickel sulphate electrolytic	Determination of: Sulphamate, chloride, sulphate and ammonium.
Nickel/iron electrolytic	Determination of: Ni^{2+}/Fe^{2+}-ratio, boric acid, Na^+, Cu^{2+}, saccharin and lauryl sulphate.
Nickel, nickel/cobalt electroless	Determination of: Hypophosphite, phosphite and phosphate, citric acid, succinic acid, Ni^{2+}/Co^{2+}-ratio.
Gold cyanide electrolytic	Determination of: $Au(CN)_2^-$, $Au(CN)_4^-$, $Co(CN)_6^-$, CN^-, chloride, phosphate, carbonate, $Fe(CN)_6^{3-}$ and $Fe(CN)_6^{4-}$.

metal ions to the metal. Some suitable reducing agents are hypophosphite, formaldehyde, hydrazine, or borohydride. As a catalyst, colloidal Pd/Sn^{2+} solution is most frequently employed. This is absorbed onto the surface of the object being plated and catalyses the deposition of a monolayer of the metal. Additional layers are deposited autocatalytically via the formation of metal hydrides.

The following reaction scheme has been proposed for the currentless plating of nickel using hypophosphite as reducing agent:

$$H_2PO_2^- + H_2O \xrightarrow[Pd/Sn^{+2}]{} HPO_3^{2-} + 2H^+ + H^-$$

$$2\,H^- + Ni^{+2} \longrightarrow Ni + H_2\uparrow \qquad (154)$$

$$2\,H_2PO_2^- + 2\,H_2O + Ni^{+2} \longrightarrow Ni + H_2\uparrow + 4\,H^+ + 2\,HPO_3^{2-}$$

It can be seen from this reaction scheme, that the rate of metal deposition is dependent on the concentration of the reducing agent. Since both nickel and hypophosphite concentrations are depleted during the process, the corresponding salts must be added to the plating solution from time to time. Furthermore, the efficiency of an electroless plating bath is limited by an increasing concentration of phosphite, formed as oxidation product from hypophosphite during the process.

With the aid of ion chromatography, the two phosphorous species can be determined simultaneously, so that the composition of the bath can be optimized. The differences in the composition of a new and an exhausted nickel plating solution can be seen from the chromatogram shown in Figure 9-13. The organic acids added as complexing agents can also be easily analyzed by ion exclusion chromatography (Figure 9-14).

If an object is to be plated with a nickel alloy, the concentration ratios of the metals in the solution is a crucial parameter. The chromatogram shown in Figure 9-15 illustrates the determination of Ni^{2+} and Co^{2+} in such a bath.

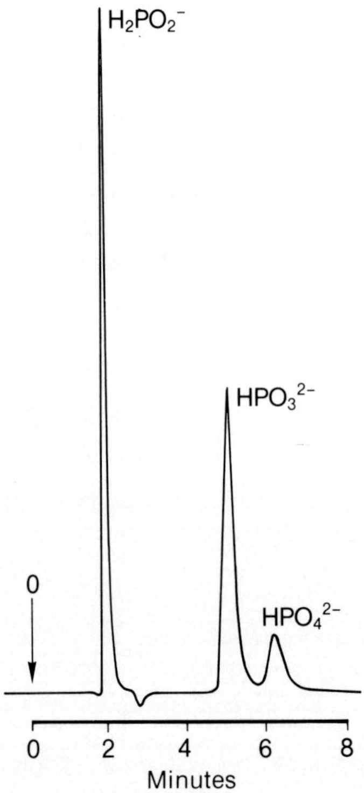

FIGURE 9-13.
The determination of hypophosphite and phosphite in a nickel/cobalt electroplating bath. **Column:** HPIC-AS3. **Eluant:** 0.003 mol/dm^3 Na_2CO_3. **Flow rate:** 2.3 cm^3/min. **Detection:** Suppressed conductivity. **Sample:** 50 μl.

FIGURE 9-14.
The determination of organic acids in a nickel/cobalt electroplating bath. **Column:** HPIC-AS1. **Eluant:** 0.001 mol/dm^3 HCl. **Flow rate:** 0.8 cm^3/min. **Detection:** Suppressed conductivity. **Sample:** 50 μl.

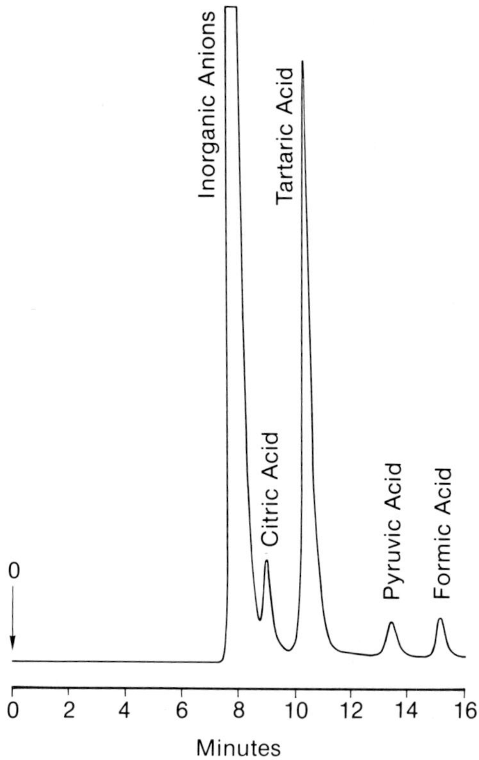

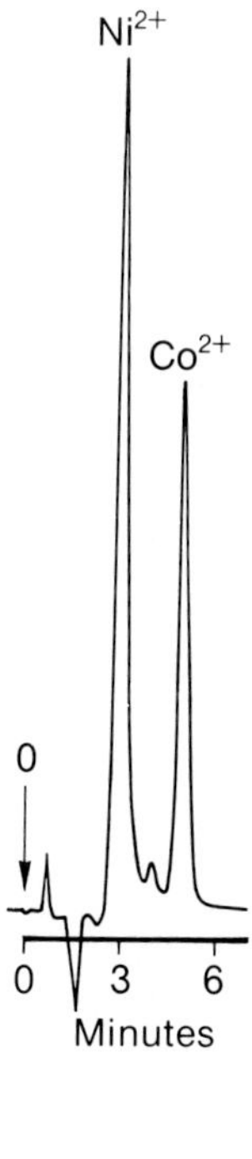

FIGURE 9-15.
A determination of the nickel/cobalt ratio in a nickel/cobalt electroplating bath. **Column:** HPIC-AS4. **Eluant:** 0.025 mol/dm^3 oxalic acid, pH 4.8 with NaOH. **Flow rate:** 1 cm^3/min. **Detection:** Postcolumn reaction with PAR and photometry at 520 nm. **Sample:** 10 μl, diluted 1:1000.

ELECTROLYTIC GOLD PLATING

In an electrolytic gold bath, the ratio of the concentrations of gold (I) and gold (III) must be continually analyzed, since the efficiency of the bath decreases as the concentration of gold (III) increases. In the presence of an excess of cyanide, both gold species Au^+ and Au^{3+} are present in the form of their complexes $Au(CN)_2^-$ and $Au(CN)_4^-$. Thus, conventional analytical methods—such as atomic absorption spectroscopy, gravimetric, and titration procedures—require considerable, time consuming sample preparation in order to decompose these complexes. Furthermore, these methods cannot easily differentiate between the two oxidation states.

Because the anionic gold complexes are stable, these species can be separated using ion pair chromatography and detected via their conductivity. Figure 9-16

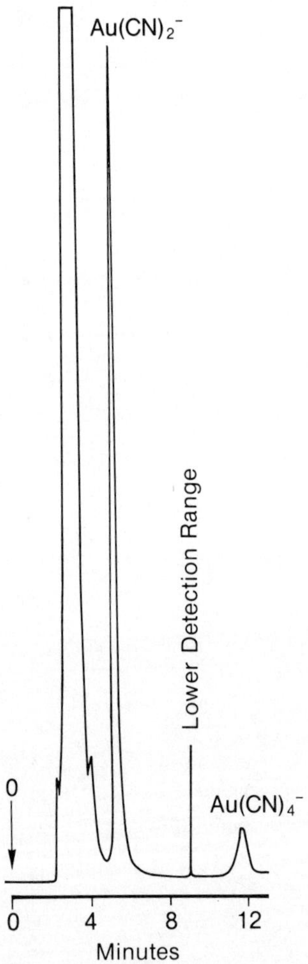

FIGURE 9-16.
The separation of gold (I) and gold (III) cyano complexes from a gold electroplating bath. **Column:** MPIC-NS1. **Eluant:** 0.002 mol/dm^3 tetrabutyl ammonium hydroxide + 0.0002 mol/dm^3 Na_2CO_3 + 40 vol.-% acetonitrile. **Flow rate:** 1 cm^3/min. **Detection:** Suppressed conductivity. **Sample:** 50 μl, diluted 1:200.

shows an example of a chromatogram obtained from a gold plating solution buffered with phosphate and citrate, simply by diluting and filtering the original sample.

Cobalt, as $CoSO_4$, is added to some plating baths as a hardener. The cyanide complex, $Co(CN)_6^{3-}$ (formed during the plating process and ineffective as hardener), can also be separated and detected in the presence of the gold cyanide complexes (see Figure 9-17). Additionally, the major components—such as phosphate and citrate buffers—as well as impurities such as chloride and iron are also easily analyzed by ion chromatographic techniques. Free cyanide can be detected down to trace levels using amperometric techniques with an Ag-working electrode.

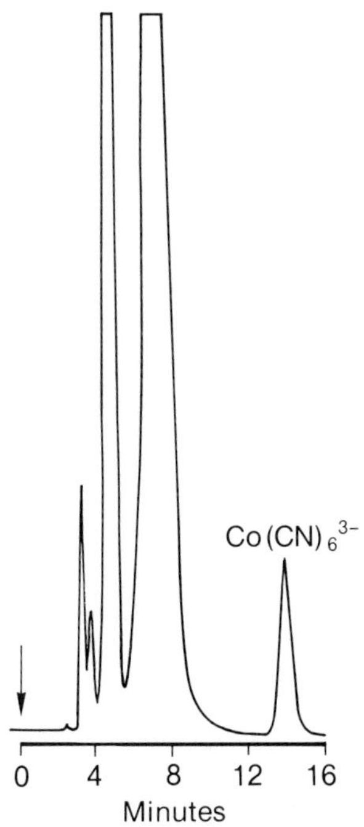

FIGURE 9-17.
A determination of the cobalt cyanide complex in a gold electroplating bath. **Column:** MPIC-NS1. **Eluant:** 0.002 mol/dm^3 tetrapropyl ammonium hydroxide + 10 vol.-% acetonitrile. **Flow rate:** 1 cm^3/min. **Detection:** Suppressed conductivity. **Sample:** 50 μl, diluted 1:400.

9.3 ION CHROMATOGRAPHY IN THE SEMICONDUCTOR INDUSTRY

One of the key problems in the semiconductor industry is the quality of the complex IC units produced which is influenced by the presence of ionic contaminants. The most important origins of these contaminants are:

- Poor quality of the water used.
- Insufficient rinsing.
- Use of impure chemicals.
- Contamination during passivation.
- Poor quality of the polymers used, etc.

The complexity, speed of development, and the diversification of this industry, have demanded the highest quality; both in the chemicals and solvents used and the analytical methods employed.

In the area of semiconductor production, ion chromatography offers the most powerful analytical method for analyzing the ionic species. Conventional methods for identifying and quantifying the ionic species responsible for corrosion on the basis of SEMI regulations (**S**emiconductor **E**quipment and **M**aterials **I**nstitute, Inc.) are labor intensive and time consuming. In comparison, it is possible to obtain complete anionic and cationic profiles within 10–15 minutes using ion chromatographic techniques. The various areas of application and typical analytical examples are summarized in Table 9-4.

The many possible applications for ion chromatography in the field of microelectronics can be emphasized using two examples.

Etching solutions—such as mixtures of hydrofluoric and nitric acids or of hydrofluoric, nitric, and acetic acids—are used to remove impurities and metal oxides from the metal surfaces. The efficiency of the etching process depends, not only on the acid composition of the etching solution, but also on the concentration of impurities in the etching solution. Figure 9-18 shows the chromatogram of a HF/HNO_3 etching solution diluted in the ratio 1:1,000 with deionized water. It should be noted that the total fluoride concentration, and not just the free hydrofluoric acid concentration, is analyzed in this way, since the metals dissolved during the etching process are present in the form of fluoride complexes (e.g., FeF_2^+). These complexes decompose in alkaline solution to form the corresponding metal hydroxides. Indirect analytical methods, such as that developed by Dulski [21], suggest that the new methods agree well with traditional wet chemical methods [22].

TABLE 9-4: Applications and typical analytical examples in the semiconductor industry.

APPLICATION	EXAMPLES	LITERATURE
Water analysis	Determination of: Inorganic mineral acids in deionized water, process, rinsing and wastewater.	[14], [15]
Etching solutions	Determination of: Main components and impurities in HF/HNO_3^- and $HF/HAc/HNO_3^-$ mixtures.	[16]
Solvents	Determination of: Chloride (direct or after extraction).	[17]
Acids	Determination of: Impurity anions like chloride, nitrate and sulphate in phosphoric acid (p.a.).	[18]
Hydroxides	Determination of: Sodium in a 46 percent KOH-solution.	[18]
Polymers	Determination of: Anionic and cationic impurities in encapsulation plastics.	[19], [20]

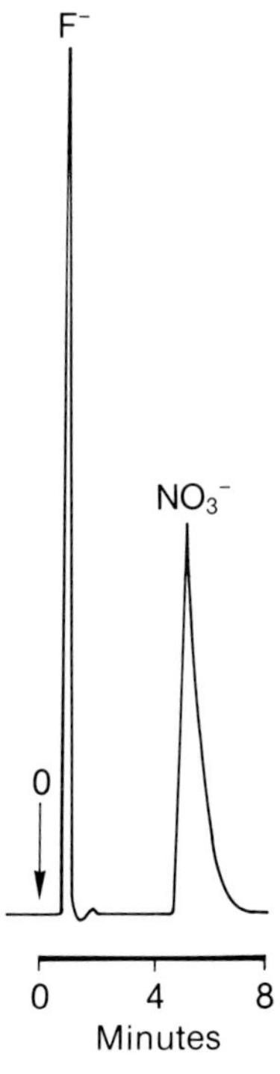

FIGURE 9-18.
Chromatogram of an HF/HNO_3 etching solution. **Column:** HPIC-AS3. **Eluant:** 0.0028 mol/dm^3 $NaHCO_3$ + 0.0022 mol/dm^3 Na_2CO_3. **Flow rate:** 2.3 cm^3/min. **Detection:** Suppressed conductivity. **Sample:** 50 μl, diluted 1:1000.

Defects in semiconductor elements often have their origin in the soldering process because the solder contains corrosive anions. After the soldering process, these anions lead to corrosion phenomena within the encapsulation casing in which the microchips are embedded. Figure 9-19 shows the chromatogram obtained from an analysis of the inorganic mineral acids from semiconductor elements. To analyze for the corrosive anions, the defective casings are broken open and extracted with 2 cm^3 deionized water. Using a direct injection technique, a chromatogram such as that shown in Figure 9-19 can be obtained. As well as the relatively high concentrations of nitrate and chloride, a number of other mineral acids and oxalic acid can be detected in low concentrations. Thus, by employing ion chromatography, the soldering process and the solder can be optimized.

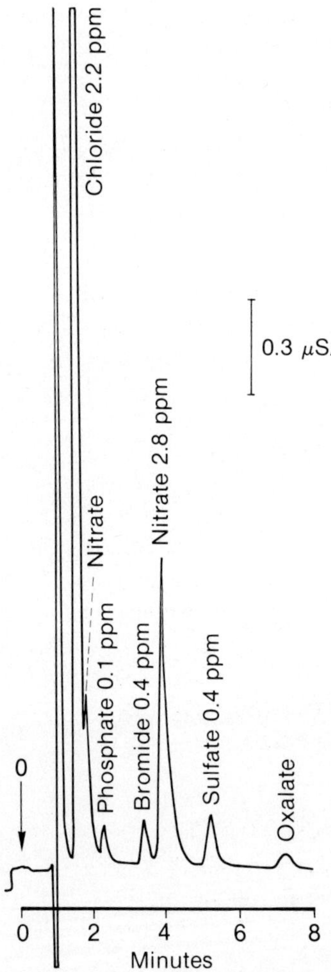

FIGURE 9-19.
An analysis of the inorganic mineral acids in a "dual in line" casing. **Column:** HPIC-AS4. **Eluant:** 0.0028 mol/dm^3 $NaHCO_3$ + 0.0022 mol/dm^3 Na_2CO_3. **Flow rate:** 2 cm^3/min. **Detection:** Suppressed conductivity. **Sample:** 50 μl.

9.4 ION CHROMATOGRAPHY IN THE FOOD AND BEVERAGE INDUSTRY

A major aim of the food and beverage industry is to provide the customers with tasty and healthy products whose nutritive value and storage quality are well defined. Production and sampling processes are automated to such an extent that only a few, most simple, quality control operations are necessary. However, in some cases, it is still necessary to carry out complex analyses to prove the quality of the products. In these cases, ion chromatography permits the rapid and precise determination of both organic and inorganic cations and anions even in complex matrices.

Ion chromatography, mainly because of the ease of sample preparation, is becoming increasingly used by both research and control laboratories in the food and beverage industry. The samples are usually extracted with deionized water and filtered before injecting them directly into the system. The reason why simple sample preparation leads to acceptable results lies, not only in the stability and robustness of the styrene/divinyl benzene resins, but also in the specificity of the detection systems.

The chromatogram shown in Figure 9-20 is a good example of the capability of ion chromatography. A weighed sample of canned spinach was extracted with deionized water, filtered, and directly injected. The inorganic anions and organic acid anions—such as tartrate and oxalate—present in the spinach could be analyzed in less than 10 minutes. From Figure 9-21, it can be seen that an analysis of acetate and chloride in mayonnaise can be accomplished in only three minutes. This permits production to be checked every 10–15 minutes and, since the analytical results are available within a very short time, changes in the composition can be rapidly made to the continuing production so that quality defects and longer production stops can be avoided. Organic acids are also used in the manufacture of cheese as preservatives or to speed up the ripening process. These acids can be easily and accurately analyzed using ion chromatography with suppressed conductivity detection rather than by measuring UV absorption or refractive indices. Food often contains components, such as carbohydrates, that interfere strongly with both of these detection methods.

For the analysis of cations, ion chromatography is particularly suitable; for example, the determination of choline in the soya products often added to baby foods. Whereas the conventional methods of choline analysis are time consuming and inaccurate, ion pair chromatographic techniques allow an accurate analysis within 20 minutes (Figure 9-22).

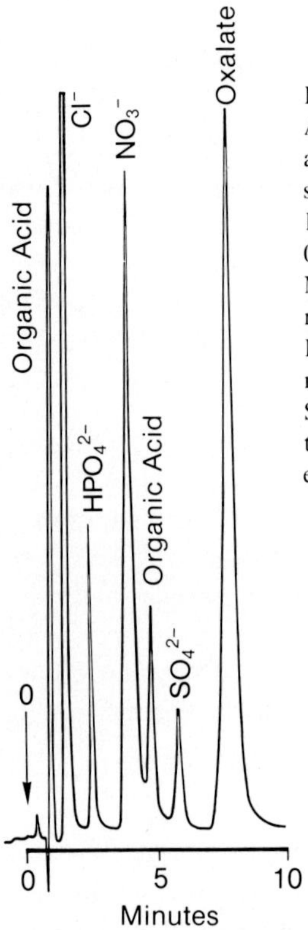

FIGURE 9-20. An analysis of the anions in canned spinach. **Column:** HPIC-AS4. **Eluant:** 0.0028 mol/dm^3 $NaHCO_3$ + 0.0022 mol/dm^3 Na_2CO_3. **Flow rate:** 2 cm^3/min. **Detection:** Suppressed conductivity. **Sample:** 50 μl extract.

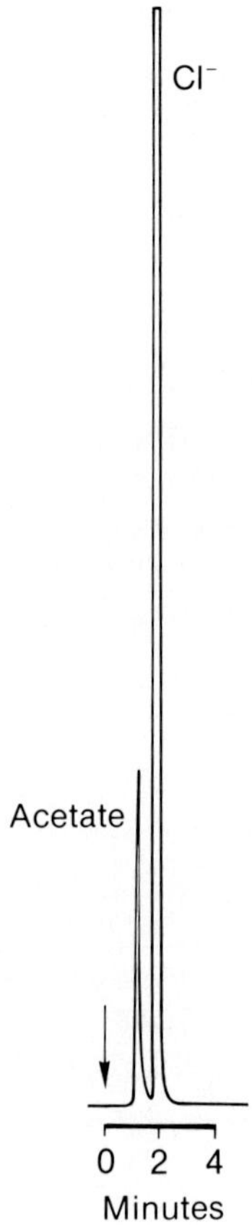

FIGURE 9-21. A determination of acetate and chloride in mayonnaise. **Column:** HPIC-AS4. **Eluant:** 0.0028 mol/dm^3 $NaHCO_3$ + 0.0022 mol/dm^3 Na_2CO_3. **Flow rate:** 2 cm^3/min. **Detection:** Suppressed conductivity. **Sample:** 50 μl extract.

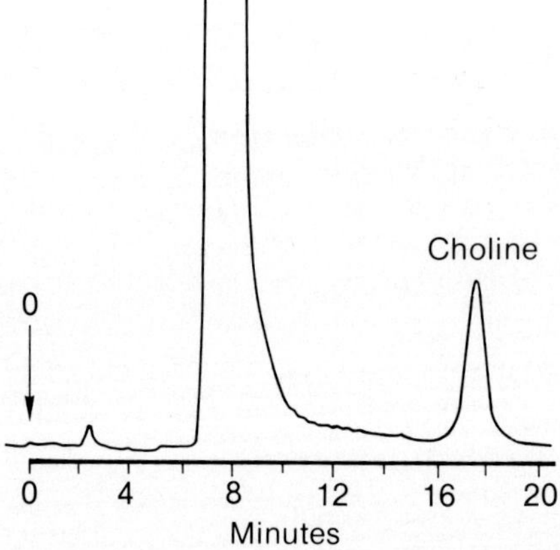

FIGURE 9-22. A determination of choline in baby food. **Column:** MPIC-NS1. **Eluant:** 0.002 mol/dm^3 + 5 vol.-% acetonitrile. **Flow rate:** 1 cm^3/min. **Detection:** Suppressed conductivity. **Sample:** 50 μl extract.

An important advantage of ion chromatographic techniques for analyzing transition metals, is that metal ions having different charges can be differentiated. Figure 9-23 shows the chromatogram of a cider vinegar. As well as iron (II), iron (III), and zinc, copper and lead could also be detected in this single run. The analysis of metals is not only important for determining compositions but can also help to identify the presence of illegal additives.

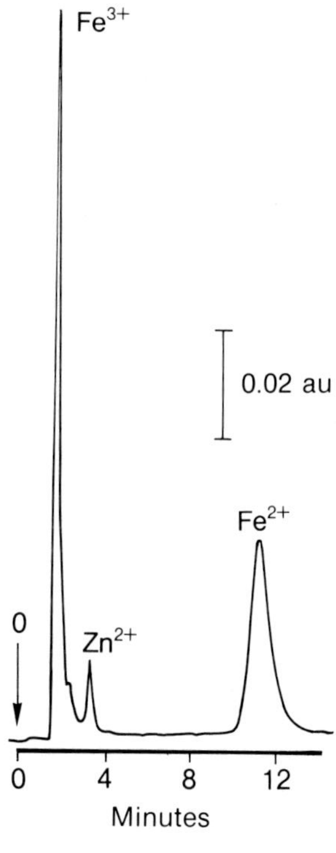

FIGURE 9-23.
A determination of the heavy metals in wine vinegar. **Column:** HPIC-CS2. **Eluant:** 0.01 mol/dm^3 citric acid + 0.0075 mol/dm^3 oxalic acid, pH 4.3 with LiOH. **Flow rate:** 1 cm^3/min. **Detection:** Postcolumn reaction with PAR and photometry at 520 nm. **Sample:** 50 μl.

A summary of the applications of ion chromatography in the food and beverage industry is given in Table 9-5.

TABLE 9-5: Applications and typical analytical examples in the food and beverage industry.

APPLICATION	EXAMPLES	LITERATURE
Milk products	Determination of: Iodide in milk, chloride and/or sodium in butter, lactate, pyruvate and citrate in cheese.	[23], [24], [25]
Meat processing	Determination of: NO_2^-/NO_3^--ratio in meat products, nitrate in the water used.	
Beverages	Determination of: Inorganic anions and cations in the used water, in sweeteners and perfumes and in the products, organic acids and carbohydrates in beer and juice.	[26], [27], [28], [29]
Condiments	Determination of: Chloride, nitrate, sodium, organic acids and heavy metals in fruit and vegetable conserves, spices, vinegar and fish.	[30], [31]
Baby food	Determination of: Choline and heavy metals.	[32]
Cereal products	Determination of: Bromate and propionate in baked products, Fe^{3+}, FE^{2+}.	[31]
Fats, oils, carbohydrates and perfumes	Determination of: Fatty acids, carbohydrates in syrup.	[33]

9.5 ION CHROMATOGRAPHY IN THE DETERGENT INDUSTRY

Detergents for both household and industrial use are very complex mixtures containing compounds of various classes of materials. The components can be divided into the following classes:

- Surfactants.
- Builders.
- Bleaches.
- Fillers and Finishing Materials.

Recent advances in ion chromatography have led to considerable progress in the analysis of detergents which, up to now, involved time consuming methods for the analysis of the various ionic species.

SURFACTANTS

The surfactants are the most important group of compounds and are contained in all detergents. Generally, these compounds are water soluble, surface-active reagents having long, hydrophobic alkyl chains and a hydrophilic functional group. Surfactants can be divided into the following classes:

- Anionic surfactants.
- Cationic surfactants.
- Nonionic surfactants.
- Amphoteric surfactants.

Those surfactants that can, at present, be analyzed by ion chromatographic techniques are summarized in Table 9-6.

The surfactants listed in Table 9-6 have been traditionally analyzed by a two-phase titration [34, 35]. This method is based on an equilibrium reaction between surfactant-color salts and anionic-cationic surfactant salts and also on the distribution of these species between chloroform and water in a two phase mixture. For the analysis of anionic surfactants, Hyamine 1622 (bis-i-butyl-phenoxyethyl-dimethyl-benzyl-ammonium chloride) has proved to be a useful titrant; whereas for cationic surfactants, lauryl sulphate is used.

A chromatographic analysis of the class of substances listed in Table 9-6 can be achieved using ion pair chromatography, coupled with a conductivity detector. The retention behavior of surface-active anions has already been discussed in Chapter 5, Section 5.5. In order to make an indubitable identification of the surfactants in a detergent, a wet chemical preseparation of the individual surfactant types is necessary. The C-chain distribution can then be determined by ion chromatography. If a crude material contains several different types of material, comparison of the chromatographic "fingerprints" can provide valuable information for the raw material control.

TABLE 9-6: Summary of those surfactants that can be analyzed by ion chromatography.

FORMULA		CHEMICAL TYPE
$R-CH_2-COONa$	$R = C_{10-16}$	Soap
$R-C_6H_4-SO_3Na$	$R = C_{11-13}$	Alkyl benzene sulphonate
$R(R)>CH\ SO_3Na$	$R + R' = C_{10-17}$	Alkane sulphonate
$R-CH_2-CH = CH-(CH_2)_n-SO_3Na$	$R = C_{10-14}$ $n = 1-3$	Alkene sulphonate
$R-CH_2-CH(OH)-(CH_2)_n-CH_2SO_3Na$	$R = C_{9-13}$ $n = 1-2$	Hydroxy alkane sulphonate
$R-CH(SO_3Na)-COOCH_3$	$R = C_{14-16}$	α-Sulpho-fatty acid methylester
$R-CH_2-O-SO_3Na$	$R = C_{11-17}$	Alkyl sulphate
$R(R)>CH-O-(C_2H_4O)_2-SO_3Na$ a)	$R' = H$ $R = C_{11-13}$ b) $R + R' = C_{10-14}$	Alkyl-ether sulphate: a) Fatty alcohol ether sulphate, b) Sec. Alkyl ether sulphate
$R_2N^+(CH_3)_2$	$R = C_{16-18}$	Di-alkyl-dimethyl ammonium chloride
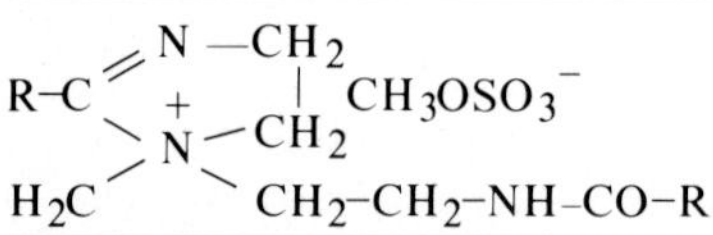	$R = C_{16-18}$	Imidazolinium salts
$C_6H_5-CH_2-\overset{+}{N}(R)-(CH_3)_2$	$R = C_{8-18}$	Alkyl-dimethyl-benzyl ammonium chloride

Cationic surfactants are not a component of common detergents, since they form water insoluble ion pairs with anionic surfactants. They are, however, often present in softener concentrates. Figure 9-24 shows the result of a chromatographic separation of didodecyl-dimethyl-ammonium chloride, an example of such a raw material.

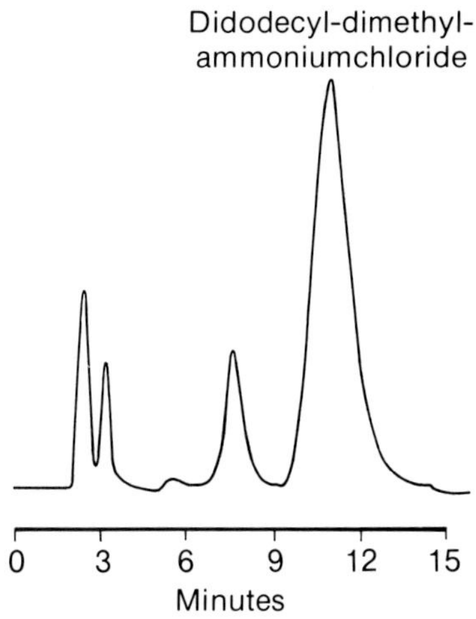

FIGURE 9-24.
An analysis of didodecyl dimethyl ammonium chloride. **Column:** MPIC-NS1. **Eluant:** 0.005 mol/dm^3 $HClO_4$ + 65 vol.-% acetonitrile. **Flow rate:** 1 cm^3/min. **Detection:** Suppressed conductivity. **Sample:** 50 μl.

BUILDERS

The builders play a key role during the washing process. They bind the calcium and magnesium ions, originating in the dirt or the textiles being washed. Among the most important builders are sodium carbonate and sodium silicate, as well as complexing agents such as sodium triphosphate and ion exchange materials such as Zeolite A.

Carbonate and silicate can be analyzed using ion exclusion chromatography. Whereas carbonate can be detected using a conductivity detector, silicate is best detected—after derivatization with sodium molybdate—photometrically at 410 nm.

Sodium triphosphate was usually analyzed using a time consuming classical ion exchange chromatographic technique and, because of its polyvalent character, was not amenable to modern ion chromatographic analysis. However, in conjunction with the recently developed anion exchange materials and specific postcolumn reactors, the analysis of sodium triphosphate—as well as orthophosphate, diphosphate, and sulphate—can now be done in a single run (see Figure 9-25). The determination of organic complexing agents—such as 1-hydroxyethane-1, 1-diphosphonic acid (HEDP), amino-tri (methylene phosphonic acid), NTA, and EDTA—can also be achieved by a similar technique [36]. Such an analysis involves simply dissolving the sample in deionized water,

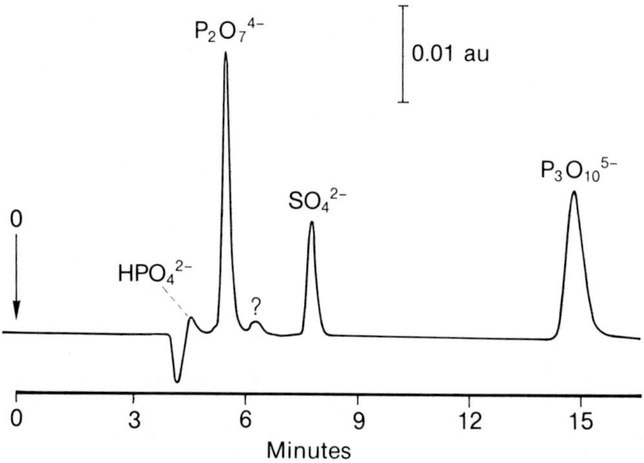

FIGURE 9-25.
An analysis of the condensed phosphates in a detergent. **Column:** HPIC-AS7. **Eluant:** 0.07 mol/dm^3 HNO^3. **Flow rate:** 0.5 cm^3/min. **Detection:** Postcolumn reaction with iron (III) nitrate and UV detection at 330 nm. **Sample:** 50 μl of a solution containing 1 g/dm^3 detergent in H_2O.

filtering, and injecting it directly into the system. As can be seen from Figure 9-26, the presence of the other detergent components do not interfere with the analysis.

BLEACHES

The most widely used bleaches are the peroxide bleaches [37] from which the active reagent, the perhydrol anion, forms from hydrogen peroxide in alkaline media according to:

$$H_2O_2 + OH^- \rightleftharpoons H_2O + HO_2^- \qquad (155)$$

The hydrogen peroxide is produced from inorganic peroxides, such as sodium perborate which is in the peroxoborate form in the crystalline state.

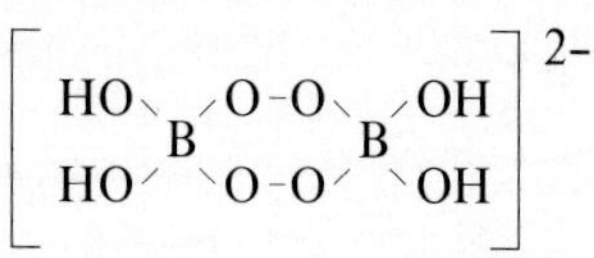

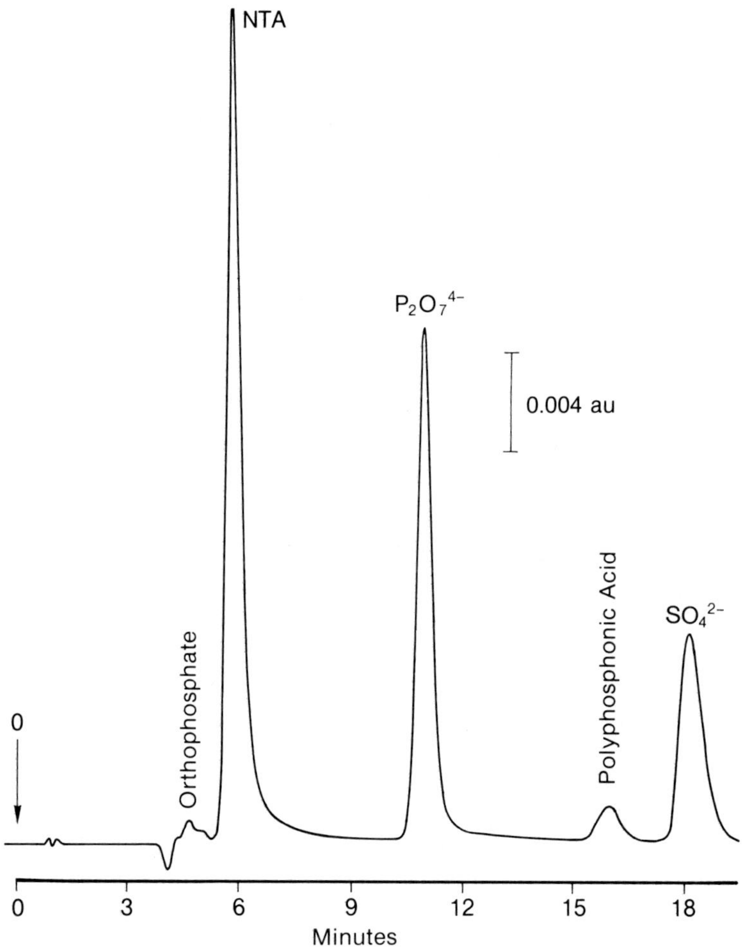

FIGURE 9-26.
An analysis of the inorganic and organic complexing agents in a detergent. **Column:** HPIC-AS7. **Eluant:** 0.03 mol/dm^3 HNO_3. **Flow rate:** 0.5 cm^3/min. **Detection:** Postcolumn reaction with iron (III) nitrate and UV detection at 330 nm. **Sample:** 50 μl of a solution containing 2 g/dm^3 detergent in H_2O.

In water, these compounds hydrolyze to yield hydrogen peroxide. In order to achieve an efficient bleaching action at temperatures less than 60° C, so called "bleach activators" are added. These activators react with hydrogen peroxide at pH's between 9 and 12 to form organic peracid intermediates that are effective bleaches at lower temperatures. Often used tetra-acetylethylene diamine (TAED) reacts in the following way:

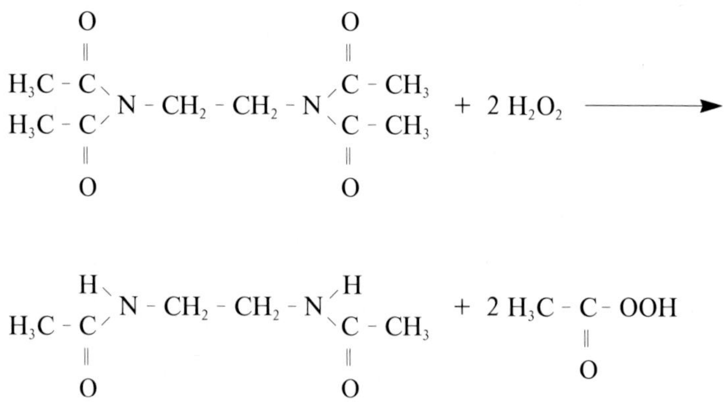

(156)

As can be seen from Figure 9-27, both the perborate and the acetic acid resulting from the decomposition of TAED can be easily and rapidly analyzed using ion chromatography. Since the separation of the individual components is achieved using ion exclusion chromatography, it is also possible to analyze for carbonate in the same run.

FILLERS AND FINISHING MATERIALS

Sodium sulphate is generally used as a filler in powdered detergents, since it leads to a powder having good flow characteristics. Sulphate can be analyzed in the presence of chloride and orthophosphate in a single anionic exchange chromatographic run (Figure 9-28). In this analysis, the orthophosphate comes from the decomposition of sodium triphosphate via the diphosphate anion. For finishing liquid detergents, sulphonated shortchain alkyl benzenes—such as toluene sulphonate or cumene sulphonate—are often used. The analysis of such components has already been discussed in Chapter 5, Section 5.5 (Figure 5-18).

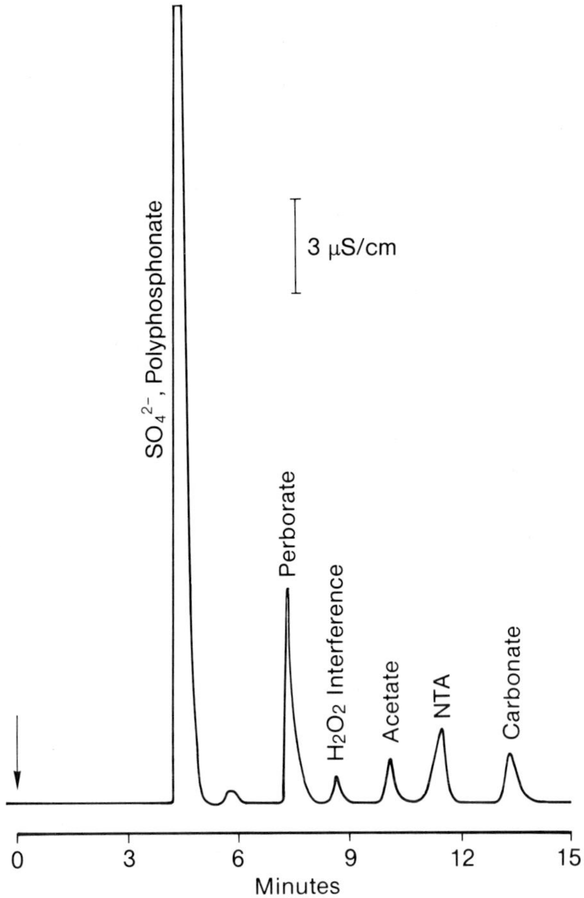

FIGURE 9-27.
An analysis of sodium perborate using ion chromatography. **Column:** HPICE-AS1. **Eluant:** 0.001 mol/dm^3 octane sulphonic acid. **Flow rate:** 1 cm^3/min. **Detection:** Suppressed conductivity. **Sample:** 50 μl of a solution containing 0.5 g/dm^3 detergent in H_2O.

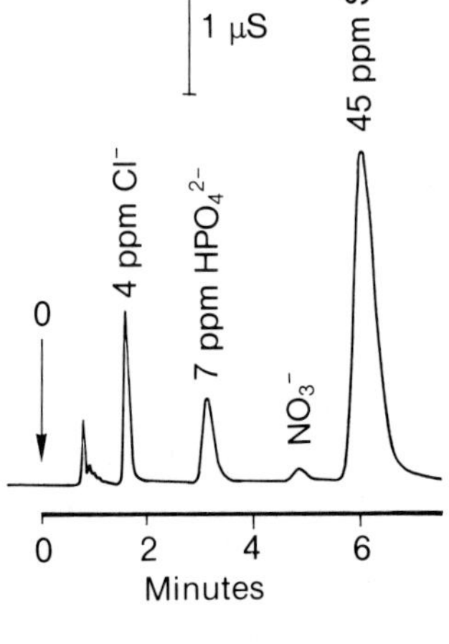

FIGURE 9-28.
An analysis of the inorganic anions in a detergent. **Column:** HPIC-AS3. **Eluant:** 0.0028 mol/dm^3 $NaHCO_3$ + 0.0022 mol/dm^3 Na_2CO_3. **Flow rate:** 2.3 cm^3/min. **Detection:** Suppressed conductivity. **Sample:** 50 μl of a solution containing 1 g/dm^3 detergent in H_2O, diluted 1:5.

9.6 ADDITIONAL AREAS OF APPLICATION

As well as in those applications described above, ion chromatography is becoming increasingly important in other analytical areas. The large number of additional areas of application, summarized in Table 9-7, cannot all be discussed in detail in this work. Therefore, one area of general interest—environmental protection—is taken as an example and discussed below.

Within the general heading of environmental protection, "acid rain" has become very important. Increasing SO_2 and NO_x emissions and the resulting acid precipitation have already threatened large areas of coniferous forests. In order to investigate the reasons for the death of the trees, it is essential to analyze the composition of the "acid rain." Ion chromatography offers the analytical chemist a rapid and selective method for determining the anions in rain water. Due to its high degree of reliability, ion chromatography has been accepted by the United States Environmental Protection Agency as the method for analyzing chloride, phosphate, nitrate, and sulphate in rain water [38].

Using the example of a galvanizing solution, the previously used and partially time consuming classical analytical methods for the determination of fluoride, hexafluorosilicate, chloride, nitrate, chlorate, and phosphate in the presence of large amounts of nickel and zinc can be substituted by fast and simple ion chromatographic technique. Since galvanizing solutions have a pH of 3.5 and are in a delicate equilibrium, it was reasonable to suspect that, after injection, the zinc phosphate would precipitate out of eluant whose pH was ca. 10. This would lead to analytical errors. However, since the samples were considerably diluted before injection, the solubility product for zinc phosphate was not exceeded.

These facts could be verified by first chromatographing the sample being analyzed using a strongly acidic cation exchange column. Appropriate comparison with untreated samples showed no statistically significant differences between the results for the treated and untreated samples. For separating the inorganic anions, an ion pair chromatographic procedure has been developed giving a baseline

TABLE 9-7: Summary of additional areas of application and appropriate analytical examples.

APPLICATION	EXAMPLES	LITERATURE
Environmental protection: Rain water, river water, aerosols, air, car exhaust fumes.	Determination of: Inorganic anions and cations, formaldehyde.	[39], [40], [41]
Medicine: Body fluids, infusion solutions, dialysis.	Determination of: Inorganic and organic acids, alkali/alkaline earth, ammonium, sugar, amino acids.	[42], [43], [44]
Paper industry: Pulping, bleaching processes, fabrication of paper, wastewater.	Determination of: Sulphur- and chloro-species, alkali and ammonium.	[45]
Chemical industry: Input- and quality control.	Determination of: Main components and impurities.	[46]
Agriculture: Fertilizers, soil extracts.	Determination of: Inorganic anions, alkali/alkaline earth, heavy metals and ammonium.	[47]
Pharmaceutical industry: Fermentation, elementary analysis, contrast materials.	Determination of: Inorganic anions and cations, organic acids.	[48], [49], [50]
Mining and ore processing: Salt mining, coal processing, phosphate coating.	Determination of: Inorganic anions, alkali/alkaline earth, metal complexes, heavy metals.	[51], [52], [53], [54], [55]
Petrochemical industry: Rinsing solutions, mineral oil, process water, environmental poisons.	Determination of: Sulphur compounds, cyanides, heavy metals, inorganic anions.	[56], [57]

resolved separation of the individual components (Figure 9-29). Thus, using ion chromatographic techniques, the analytical problems of determining the ionic components of phosphate coating solutions can be solved more effectively.

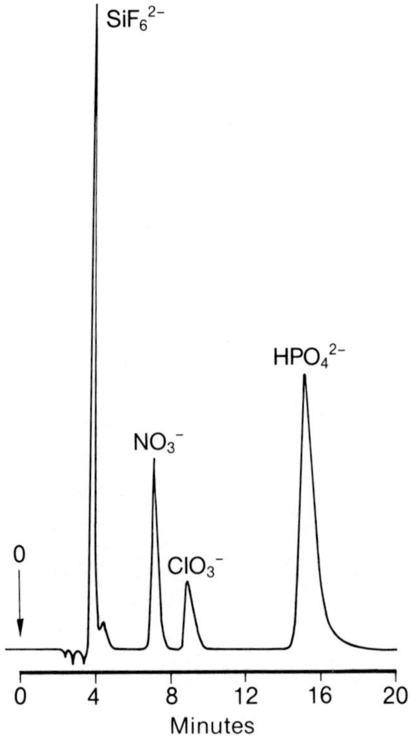

FIGURE 9-29.
A separation of the inorganic anions from a galvanizing solution. **Column:** MPIC-NS1. **Eluant:** 0.002 mol/dm^3 tetrabutyl ammonium hydroxide + 0.001 mol/dm^3 Na_2CO_3 + 9.5 vol.-% acetonitrile. **Flow rate:** 1 cm^3/min. **Detection:** Suppressed conductivity. **Sample:** 50 μl, diluted 1:25.

9.7 SAMPLE PREPARATION AND MATRIX PROBLEMS

As with all chromatographic techniques, effects occur during ion chromatography influencing the efficiency of the analysis. Among such effects are:

- Decrease in the resolution and exchange capacity.
- Increasing backpressure from the column.
- Baseline changes.

Although these phenomena cannot be completely eliminated, the frequency of their occurrence can be reduced by suitable sample preparation and appropriate care of the columns.

A decrease in the separation performance of a column can usually be attributed to a blocking of the exchange functions by substances having a relatively high affinity for the stationary phase. Thus, analytical columns should be rinsed from time to time with a suitable eluant to remove such contaminants [58].

One of the most effective ways of protecting an analytical column is to use a pre-column (guard column), packed with the same stationary phase as that in the analytical column. Ions having a high affinity for the stationary phase are retained by the precolumn and a "poisoning" of the analytical column is avoided. Since the capacity of the precolumn is also limited, it, too, must be rinsed from time to time to remain effective. A typical example of such effects is the analysis of inorganic mineral acids in fruit juices containing large amounts of citric acid. The citric acid is retained on the precolumn, but can be removed by injecting a concentrated carbonate solution.

With complex matrices, such as blood plasma and urine [42], even a precolumn cannot provide adequate protection of the analytical column from possible contamination. Such samples contain proteins that have such a high affinity for the stationary phase that cleaning becomes problematic. Therefore, it is recommended that proteins be precipitated (with acetonitrile) from such mixtures before analysis. In order to keep the myriad of organic components of such mixtures away from the analytical column, an extraction cartridge—containing a chemically bound reversed phase (Baker, Millipore)—can be used. Nonionic organic materials are retained by the packing in the cartridge that can be thrown away after use.

Extraction cartridges of this type are also used during the analysis of soil extracts [47] to protect the analytical column from humus materials. Since humus contains high molecular weight polyhydroxycarboxylic acids having a high affinity for the stationary phase, direct injection of such materials would lead to a rapid, total loss of separation capacity. Should this occur, the analytical column must be subjected to a time consuming cleaning process which itself has a negative influence on the separation capacity of a column.

LITERATURE

1. T. Darimont, G. Schulze, and M. Sonneborn, Fresenius Z. Anal. Chem. **314**, 383-85 (1983).
2. G. Schwedt, "Ionen-Chromatographie (IC)—die "High Performance"—LC für Anionen und Kationen," Labor-Praxis Jan./Feb. 1984.
3. R. Merriweather, "Environmental Water Analysis by Ion Chromatography," Union Carbide, Analytical Chemistry Dept., Oak Ridge, Tennessee.
4. Dionex Corp. Sunnyvale, Technical Note Nos. 5 and 8.
5. G. Resch and E. Grünschläger, VGB Kraftwerkstechnik 62, 9 127-132 (1982).
6. Dionex GmbH Weiterstadt, Application Report No. 11/82/1.
7. Dionex GmbH Weiterstadt, Application Report No. 7/84/4.
8. J. Weib and M. Göbl, Fresenius Z. Anal. Chem. in press.
9. Dionex GmbH Weiterstadt, Application Report No. 7/84/5.
10. K. Haak and R. George, "The Use of Ion Chromatography in the Electroplating Industry," Plating and Surface Finishing, Sept. 1983.
11. Dionex Corp. Sunnyvale, Application Report No. 40.
12. Dionex Corp. Sunnyvale, Application Report No. 49.

13. Dionex Corp. Sunnyvale, Application Report No. 50.
14. Dionex Corp. Sunnyvale, Application Report No. 34.
15. G.P. Simon and C. Calmon, Solid State Technology, pp 21-30 (1968).
16. Dionex Corp. Sunnyvale, Application Report No. 28.
17. Dionex Corp. Sunnyvale, Application Report No. 32.
18. Dionex Corp. Sunnyvale, Application Report No. 35.
19. Dionex Corp. Sunnyvale, Application Report No. 33.
20. Dionex GmbH Weiterstadt, Application Report No. 3/83/5.
21. T.R. Dulski, Anal. Chem. **51**, 1439-43 (1979).
22. J.P. McKaveney, Anal. Chem. **40**, 1276-79 (1968).
23. Dionex Corp. Sunnyvale, Application Report No. 9.
24. Dionex Corp. Sunnyvale, Application Report No. 37.
25. L. Angers and K. Haak, "Sodium Determination in Food," Dionex Corp. Sunnyvale, Dept. for Method Development.
26. Dionex Corp. Sunnyvale, Application Report No. 19.
27. Dionex Corp. Sunnyvale, Application Report No. 21.
28. Dionex Corp. Sunnyvale, Application Report No. 25.
29. Dionex Corp. Sunnyvale, Application Report No. 46.
30. P. Edwards, Food Technology, pp 53-56, June 1983.
31. D. Eubanks, A.W. Fitchett, and L. Angers, "Trace Metal in Food by Ion Chromatography Using Post Column Reaction Techniques," FACSS-Meeting, Philadelphia, PA (1982).
32. Dionex Corp. Sunnyvale, Application Report No. 38.
33. Dionex Corp. Sunnyvale, Application Report No. 45.
34. G.F. Longman, "The Analysis of Detergents and Detergent Products," John Wiley & Sons, London 1975.
35. M.J. Rosen and H.A. Goldsmith, "Systematic Analysis of Surface Active Agents," Second Edition, John Wiley & Sons, London 1972.
36. A.W. Fitchett and A. Woodruff, "Determination of Polyvalent Anions by IC," LC Vol. 1 (1968) No. 1.
37. Ullmanns Encyklopädie der technischen Chemie, 4. Auflage, Bd. 24, p 97.
38. United States Environmental Protection Agency, Cincinnati, Ohio, EPA-600-4-84-017, March 1984.
39. E. Sawicki, J.D. Mulik, and E. Wittgenstein, "Ion Chromatographic Analysis of Environmental Pollutants," Vol. 1 and 2, Ann Arbor Science 1978/79.
40. Dionex Corp. Sunnyvale, Application Report No. 24.
41. Dionex Corp. Sunnyvale, Application Report No. 31.
42. Dionex GmbH Weiterstadt, Application Report No. 11/82/2.
43. W.G. Robertson, D.S. Scurr, A. Smith, and R.L. Orwell, Clinica Chimica Acta, 00 (1982) CCA 2329.
44. C.J. Mahle and M. Menon, J. of Urology **127**, 159-62 (1982).
45. G.O. Franklin and A.W. Fitchett, "Fast Chemical Characterization of Pulping and Bleaching Process by Ion Chromatography," Pulp & Paper Canady 83, No. 10 (1982).
46. Dionex GmbH Weiterstadt, Application Report No. 8/82/1.
47. Dionex GmbH Weiterstadt, Application Report No. 5/83/1.
48. R.D. Rocklin and E.L. Johnson, "Analysis of Fermentation Broths by Ion Chromatography," submitted to publication.
49. T.L. Hassinger and B.A. Olson, "Analysis of Fermentation Broths and Raw Materials Utilizing Ion Chromatographic Techniques," submitted for publication.
50. Dionex GmbH Weiterstadt, Application Report No. 11/82/3.
51. Dionex Corp. Sunnyvale, Application Report No. 40.
52. Dionex Corp. Sunnyvale, Application Report No. 41.
53. Dionex GmbH Weiterstadt, Application Report No. 8/82/2.

54. Dionex GmbH Weiterstadt, Application Report No. 3/83/1.
55. Dionex GmbH Weiterstadt, Application Report no. 8/83/2.
56. R.D. Rocklin and E.L. Johnson, Anal. Chem. **55**, 4-7 (1983).
57. P. Viswanadham, D.R. Snick, J.J. Pisney, and W.F. Dilworth, Anal. Chem. **54**, 2431-33 (1982).
58. Dionex Corp. Sunnyvale, Technical Note No. 2R.

10 Methods Related to Ion Chromatography

As a recent alternative to the method originally described by Small (Chapter 1, Section 1.3), separation and detection systems that, in principle, do not utilize a suppressor system have been developed for the analysis of inorganic ions. In the past, this type of chromatography was known as "single column ion chromatography." This designation is, however, a misnomer; since the suppressor system in connection with conductivity detection is not part of the separation system, but serves simply to suppress the background conductivity of the eluant. Thus, an alternative term for this type of chromatography has been introduced: "non-suppressed ion chromatography."

The aim of this last chapter is to summarize the separation and detection systems developed as alternatives to those based on Small's original concept and, in some cases, to compare the alternative methods. The discussion will be limited to publications describing analyses of inorganic anions, since these are much more numerous than those concerned with the corresponding chromatographic analysis of cations.

10.1 SEPARATION METHODS

The separation techniques used for the analysis of inorganic anions can be classified according to Figure 10-1 [1]. Thus, separations taking place via ion exchange on an ion exchange material with bound charges are differentiated from those separations occurring via ion pairs on nonpolar stationary phases.

SEPARATIONS ON ION EXCHANGE MATERIALS HAVING HIGH CAPACITIES

Commercially available ion exchange materials—on either a silica or a polystyrene/divinyl benzene basis—generally have high exchange capacities in the

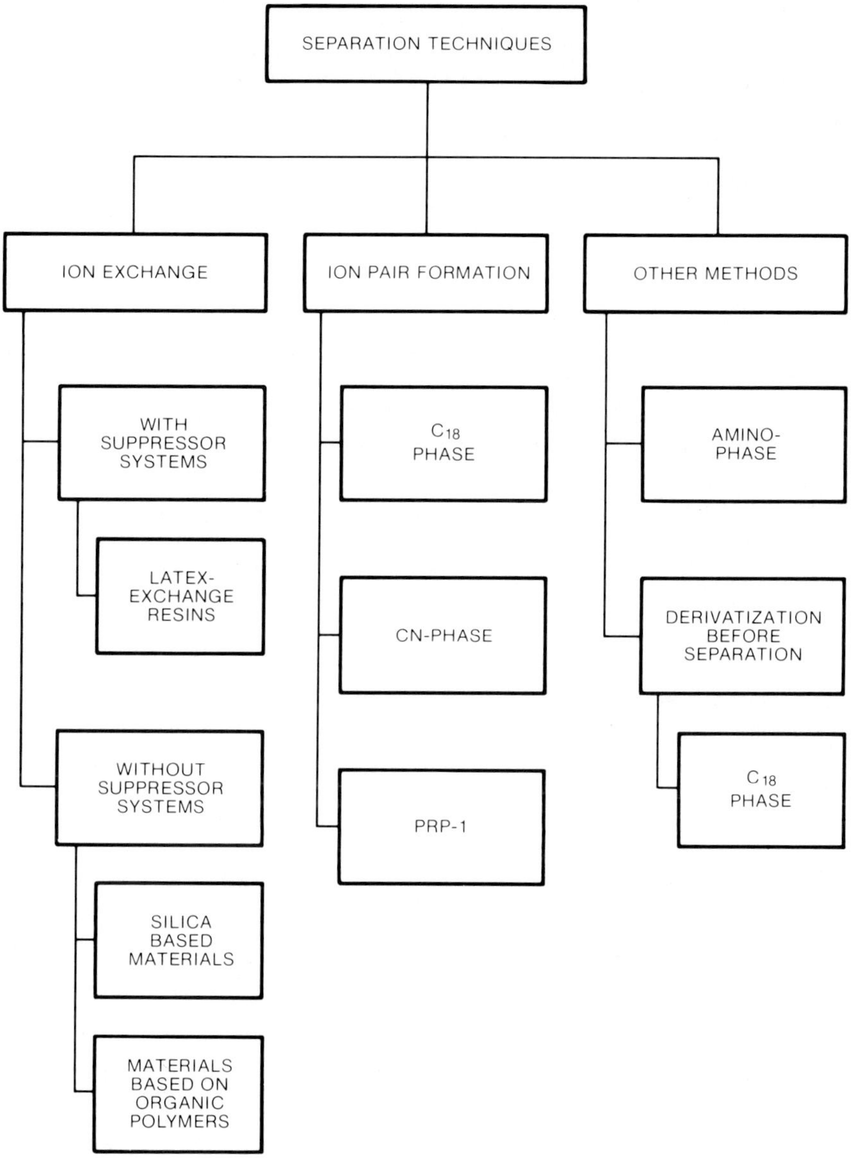

FIGURE 10-1.
Schematic representation of the various separation methods available for the analysis of inorganic anions.

range of 1 mequiv./g. Such materials are not widely used because, in order to elute inorganic ions, eluants with high ionic strengths must be employed. This excludes the possibility of using conductivity as a method of detection for the solute ions. For this reason, high capacity ion exchange materials are only used in conjunction with a photometric detector (with or without postcolumn derivatization) or amperometric detection.

For example, nitrite and nitrate can be separated on pellicular ion exchange materials based on silica [2], organic polymers [3], or cellulose [4], and detected directly with a UV detector. Recent developments towards an indirect method of detection [5,6] show that, in higher concentration ranges (> 1 ppm), the refractive index is a viable alternative to conductivity as a detection method.

SEPARATIONS ON ION EXCHANGE MATERIALS HAVING LOW CAPACITIES AND BASED ON ORGANIC POLYMERS

In the autumn of 1978, Fritz et al. [7,8] introduced a chromatographic system for analyzing inorganic anions. In this system, the analytical column was connected *directly* to a conductivity cell. A prerequisite for the functioning of such a set-up is that the anion exchange resin used has a low capacity, so that solutes can be eluted using eluants of low ionic strength. In addition, the eluant used must have a low inherent conductivity if a sensitive conductivity detection of the solute species is to be possible.

Fritz et al. fulfilled these conditions by developing a series of macroporous anion exchange resins having low capacities (0.007 to 0.04 mequiv./g) with highly crosslinked polystyrene particles (XAD-1, Rohm and Haas, Philadelphia, USA) as substrate. The substrate was initially chloromethylated and, then, its surface aminated using a solution of triethylamine in methanol. As eluant, Fritz et al. employed the sodium, potassium, or ammonium salts of benzoic, phthalic, or o-sulphobenzoic acid in concentrations between $1 \cdot 10^{-4}$ and $5 \cdot 10^{-4}$ mol/dm^3.

Experiments designed to investigate the dependence of the retention times of inorganic anions on the capacity of the resin [9] have shown that the distribution coefficient of the anions examined decreases as the exchange capacity of the resin decreases. In contrast, the selectivity coefficients—most important for the separation—remain constant within the range of capacities studied (0.04 to 1.46 mequiv./g).

Eluants

The choice of eluant is the subject of a large number of studies described in the literature [7-10]. Most frequently used have been the benzoates, phthalates, and o-sulphobenzoates; since these have a strong enough affinity for the surface aminated anion exchange resins while also having low molar conductivities. When using aromatic acids as eluants, the pH of the solution must be adjusted to between 4 and 7; since the pH determines the degree of dissociation of these acids and, thus, the retention behavior of species being analyzed. In any case, the acids must be essentially neutralized in order to minimize the concentration of the strongly conducting hydronium ions. The conductivity of the eluants used by Fritz et al. in their initial work [9] are listed in Table 10-1.

TABLE 10-1: The background conductivity due to those eluants used by Fritz et al [9].

ELUANT	CONCENTRATION	VALUE	CONDUCTIVITY
Potassimum benzoate	6.5×10^{-4} mol/dm^3	4.6 pH	65.9 μS/cm
Potassimum hydrogen-phthalate	5.0×10^{-4} mol/dm^3	4.4 pH	74.3 μS/cm
	5.0×10^{-4} mol/dm^3	6.2 pH	112 μS/cm
	6.5×10^{-4} mol/dm^3	4.4 pH	90.5 μS/cm
	6.5×10^{-4} mol/dm^3	6.2 pH	158 μS/cm
Ammonium-o-sulpho-benzoate	5.0×10^{-4} mol/dm^3	5.8 pH	132 μS/cm

Due to these eluants, the background conductivity is much higher than that resulting from carbonate/hydrogen carbonate based systems used in conjunction with a suppressor system (20 μS/cm). The higher background conductivities (between 60 and 160 μS/cm) raise the detection limits and restrict the linear dynamic range of conductivity detection. However, more recent research by Gjerde and Fritz [10] suggests that more exact neutralization of the eluant results in considerably lower conductivities.

Problems arise when adjusting the pH of the eluant using commercially available pH probes. Such probes normally consist of an Ag/AgCl reference electrode, a saturated KCl solution as base electrolyte, and a porous sinter as salt bridge. Thus, contamination of the eluant with chloride, subsequently leading to erroneous chloride analyses [11], is possible. Separation of the reference electrode from direct contact with the eluant can circumvent this problem.

Using aromatic carboxylic acids as eluant, the separation of inorganic anions is possible. Examples of such anions are: fluoride, chloride, bromide, iodide, nitrite, nitrate, sulphate, thiocyanate, and thiosulphate. However, under optimum conditions, only five anions can be separated in a single run. Sodium or potassium benzoate is preferred for separation of monovalent anions. The affinity of benzoate for the stationary phase is similar to that of hydrogen carbonate. For eluting divalent species, the corresponding salts of phthalic acid have to be used since these salts have a stronger affinity for the stationary phase. Increasing the concentration of benzoate is not possible due to the resulting, higher conductivity making sensitive conductivity detection impossible.

A typical chromatogram obtained using a surface aminated XAD-1 resin for the separation of fluoride, chloride, and bromide is shown in Figure 10-2. From the literature describing the use of such materials, it is clear that they have very low chromatographic efficiencies (ca. 1,000 to 2,000 theoretical plates per meter). The positive signal in Figure 10-2, (which corresponds to the dead volume), is characteristic of ion chromatographic systems without suppressor units and represents the cations of the sample being analyzed that are not retained by the stationary phase. The cations travel through the column and into the detector as a single band with the same speed as the eluant and associate with the anions of the eluant. If the equivalent conductivity of these salts is larger than that of the eluant, a positive peak results. If, on the other hand, these salts have a lower equivalent conductivity than the eluant, a negative peak results that can lead to problems in evaluating peaks due to ions with short retention times.

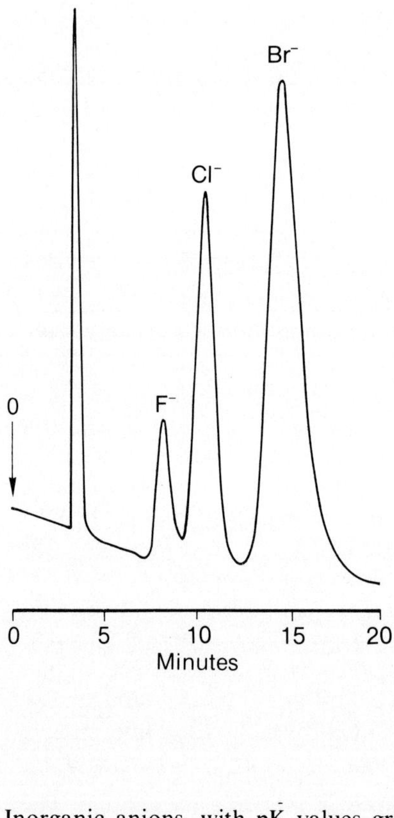

FIGURE 10-2.
The separation of fluoride, chloride, and bromide on a surface aminated XAD-1 resin. **Column:** 1000 × 2 mm ID XAD-1. **Exchange capacity:** 0.04 mequiv./g. **Eluant:** 6.5×10^{-4} mol/dm^3 potassium benzoate (pH 4.6). **Flow rate:** 2 cm^3/min. **Detection:** Direct conductivity. **Sample:** 50 μl, containing 4.8 ppm fluoride, 5.1 ppm chloride and 26 ppm bromide. (Refer to reference [9] of the Literature.)

Inorganic anions, with $p\bar{K}$ values greater than 7, can be eluted using sodium hydroxide or sodium phenate and detected directly via their conductivity. However, in systems having no suppressor, such compounds are detected as negative signals since their equivalent conductivity is less than that of the eluant.

Experimental Parameters Affecting Retention Times

Theoretical considerations concerning the retention behavior of inorganic anions in ion chromatographic systems not having suppressor systems, have recently been published by Jenke and Pagenkopf [12]. In addition to the pH, the nature and concentration of the organic acid in the eluant are the determining parameters. By varying the pH and keeping the concentration of the acid constant, and by varying the concentration of the acid while keeping the pH constant, the chromatographic conditions can be optimized [13].

A plot of the logarithm of the net retention times against the logarithm of the eluant concentration gives a straight line for each of the various inorganic anions [14]. The slopes of the lines are dependent on the nature of the eluant and the solute ion S^{y-}. This effect is demonstrated by the results obtained by Small as shown in Figure 10-3. Particularly worth noting are the coelution of some ions and the reversing of elution sequences under some conditions. Mathematically, the retention behavior of the inorganic anions can be described by:

$$\log (V_E - V_V) \propto - m \cdot \log [E^{x-}] \tag{157}$$

$V_E - V_V$: Net retention volume.
$[E^{X-}]$: Concentration of the eluant.

Alternatively, Equation 157 can be written in the form [15]:

$$\log (V_E - V_V) = \text{constant} - \frac{y}{x} \cdot \log [E^{x-}] \tag{158}$$

y: Charge of the solute ion.
x: Charge of the eluant ion.

From Equation 158, the slope of the straight line is proportional to the quotient of the charge of the solute and eluant ions. If the quotients y/x are calculated and compared (see Table 10-2) with those values measured by Small, the agreement is very good. Assuming that the charges of the solute and eluant are known, then it is only necessary to know the retention time for the ion at a *single* eluant concentration in order to be able to predict its retention at other eluant concentrations. The dependence of the slope of the line on the ratio of the charge numbers also explains the reverses in the elution sequences. On the basis of this information, the coelution of ions can be avoided by appropriate choice of chromatographic conditions.

Surface aminated macroporous resins are not efficient enough for most chromatographic applications and can, thus, only be used when a high resolution is unnecessary [16].

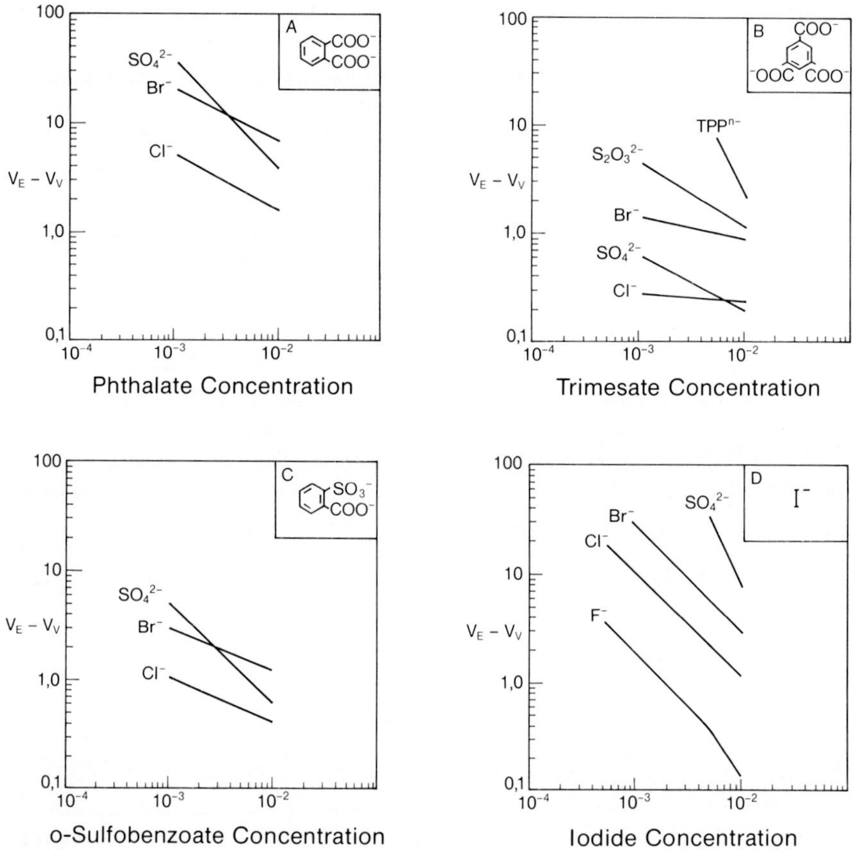

FIGURE 10-3.
The dependence of the net retention volume of various inorganic anions on the concentration of (a) phthalate (b) trimesate (c) o-sulphobenzoate and (d) iodide. **Column:** 250 × 2.8 mm ID SAR-40-0.6. (Refer to reference [14] of the Literature.)

TABLE 10-2: A comparison of the theoretical and measured values for the quotient y/x [14].

ELUANT ION E^{x-}	SOLUTE ION S^{y-}	$y/x_{theor.}$	$y/x_{meas.}$
I^-	Br^-	1	1.00
I^-	SO_4^{2-}	2	2.00
Trimesate^{3-}	Br^-	0.33	0.21
Trimesate^{3-}	SO_4^{2-}, $S_2O_3^{2-}$	0.66	0.60
Phthalate^{2-}	Br^-	0.5	0.47
Phthalate^{2-}	SO_4^{2-}	1	0.98

Highly efficient anion exchange materials (e.g., TSK gel IC-PW) derived from organic polymers have recently been introduced by Toyo Soda (Tokyo, Japan). These materials are composed of a polyacrylate with an exchange capacity of 0.003 mequiv./g having particle diameters of ca. 10 μm. The chromatographic efficiency of such a column having the dimensions of 50 mm × 4.6 mm ID is some 20,000 theoretical plates per meter, with respect to the nitrate ion. Polyacrylate resins, however, are only stable between pH 1 and pH 11. Using an eluant consisting of potassium gluconate and a borate buffer (see Figure 10-4), it is possible to analyze up to seven anions in a single run. For separating monovalent anions on TSK gel IC-PW, potassium hydroxide can also be used as eluant [17].

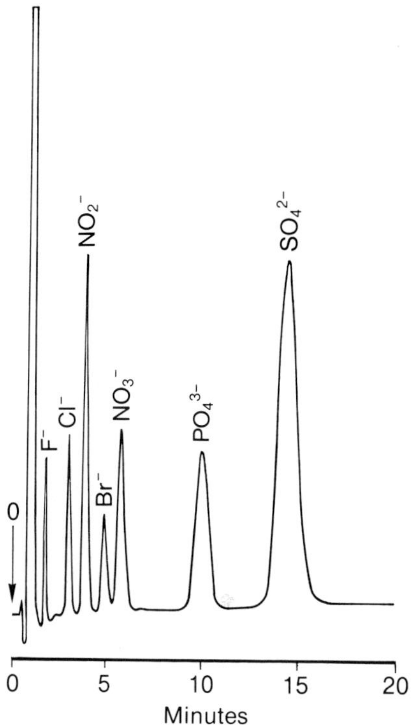

FIGURE 10-4.
The separation of inorganic ions on a TSK gel IC-PW ion exchange column. **Eluant:** 0.0013 mol/dm^3 $Na_2B_4O_7$ × 0.058 mol/dm^3 H_3BO_3 × 0.0014 mol/dm^3 potassium gluconate (pH 8.5) + 12 vol.-% acetonitrile. **Flow rate:** 1.2 cm^3/min. **Detection:** Direct conductivity. **Sample:** 100 μl. **Sample concentration:** 5–40 ppm. [From P. R. Haddad and A. L. Heckenberg, J. Chromatogr., **300,** 357 (1984).]

SEPARATIONS ON ION EXCHANGERS OF LOW CAPACITY BASED ON SILICA

Parallel to the development of organic polymers as substrates for anioin exchange materials having low capacities, novel silica-based materials [18, 19] have been introduced. The aim of this work is the development of exchange materials having low capacities enabling the use of eluants having low equivalent conductivities so that suppressor systems would become unnecessary. The synthesis of such materials (with capacities between 0.1 and 0.3 mequiv./g) is achieved by binding quaternary ammonium functions onto very small (6–10 μm) silica particles. Organic acids such as phthalic, benzoic, and salicylic acid have proved particularly suitable as eluants. Most frequently used is o-phthalic acid in the concentration range $1 \times 10^{-3} - 1 \times 10^{-2}$ mol/dm^3. By adjusting the pH of the mobile phase to between 3.5 and 6, it is possible to adjust the ionic strength and, thereby, to vary the chromatographic characteristics of the solute ions. Anion exchange materials based on silica are available, for example, from Vydac (The Separation Group, Hesparia, CA, USA), Wescan (Santa Clara, CA, USA) and Toyo Soda (Tokyo, Japan), Although these materials are characterized by relatively high chromatographic efficiencies (15,000–20,000 thoeretical plates per meter [20]), the choice of eluant is very limited. This results from the narrow pH range (pH 2-7) within which silica remains stable. Figures 10-5 and 10-6 show two typical examples of separations using materials of this kind.

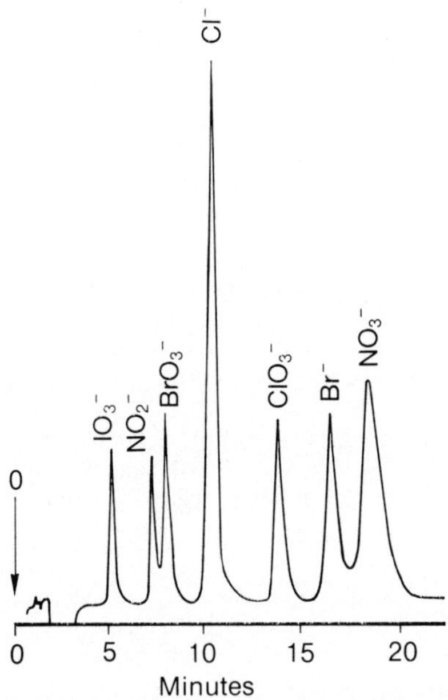

FIGURE 10-5.
The separation of inorganic anions on a low capacity anion exchange resin based on silica. **Column:** 50 × 4 mm ID TSK gel IEX-520. **Eluant:** 0.001 mol/dm^3 tartaric acid (pH 3.2). **Flow rate:** 1.5 cm^3/min. **Detection:** Direct conductivity. **Sample:** 100 μl. **Solute concentration:** 100 ppm. (Refer to reference [20] of the Literature.)

It is worth noting the behavior of transition and heavy metals in ion chromatographic systems without a suppressor (studied by Jenke and Pagenkopf [21]). Metal cations—such as Cu^{2+}, Pb^{2+}, and Zn^{2+}—can negatively influence the stability and performance of columns containing silica-based exchange materials. The free silanol groups, present in such materials, interact with these metal cations, so that they are retained on the stationary phase. Thus, the metal cations—such as lead, zinc, and copper—coelute with anions—such as chloride, bromide, and nitrate—leading to erroneous analytical results.

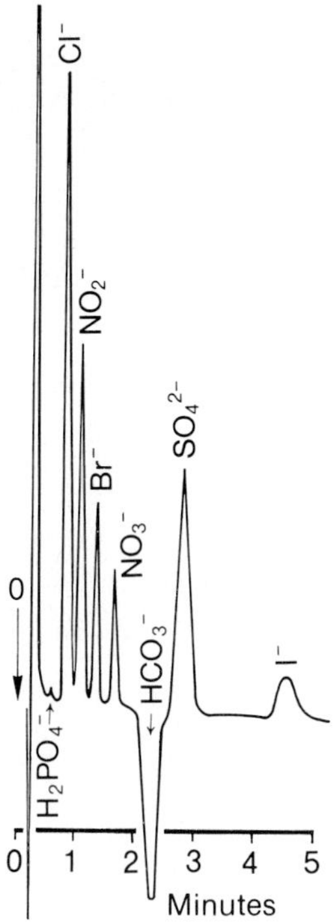

FIGURE 10-6.
The separation of inorganic anions on a low capacity anion exchange column based on silica. **Column:** 100 × 4.6 mm ID Wescan HS anion exchanger. **Eluant:** 0.004 mol/dm^3 potassium hydrogen phthalate (pH 3.9). **Flow rate:** 4 cm^3/min. **Detection:** Direct conductivity. **Sample:** 100 μl. **Solute concentration:** Not given. [From Th. Jupille, D. Burge and D. Togami, Chromatographia **242,** 177 (1982).]

Low capacity anion exchange materials based on silica lead to the same limitations on the choice of eluant as do the analogous polymer-based materials. Although the chromatographic efficiency of the materials is generally very high, they can only be used in a limited pH range, restricting the type of sample that can be analyzed using such materials.

ION PAIR CHROMATOGRAPHIC SEPARATIONS

Mechanistic and practical aspects of the ion pair chromatography on organic polymer substrates, such as polystyrene-divinyl benzene, have already been discussed in detail in Chapter 5. Compared to ion exchange chromatography, the advantage of this method lies in the adaptability of such systems to individual separation problems. This flexibility results from the large number of experimental parameters controlling the retention of the solutes. An additional advantage of ion pair chromatography is the better exploitation of the high chromatographic efficiency of the stationary phases used in comparison to the conventional exchange materials on silica and organic polymer substrates.

At this time, the highest chromatographic efficiency for the separation of inorganic anions that can be achieved is obtained using chemically bound reversed phases on silica substrates [22-24]. Additionally, ion pair chromatography has also been carried out on chemically bound cyanophases [25, 26]. Typical results from such separations are shown in Figures 10-7 and 10-8. From the literature [23, 25, 27], it can be seen that a differentiation is made when using ion pair reagents between a dynamic and a "permanent" coating of the stationary phase.

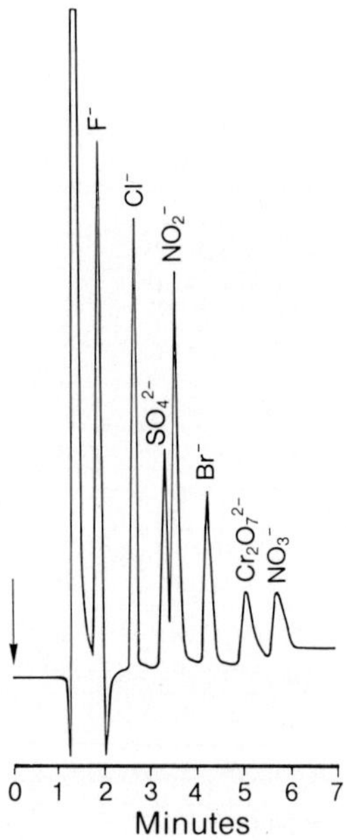

FIGURE 10-7.
Ion pair chromatographic separation of inorganic anions on a chemically bound reversed phase. **Column:** 250 × 4.4 mm ID LiChrosorb RP-18 (10 μm). **Eluant:** 0.002 mol/dm^3 TBAOH + 0.05 mol/dm^3 phosphate buffer (pH 6.7). **Flow rate:** 2 cm^3/min. **Detection:** Direct conductivity. **Sample:** 20 μl containing 1000 ppm fluoride, chloride, sulphate, nitrite, bromide, dichromate and nitrate. (Refer to reference [22] of the Literature.)

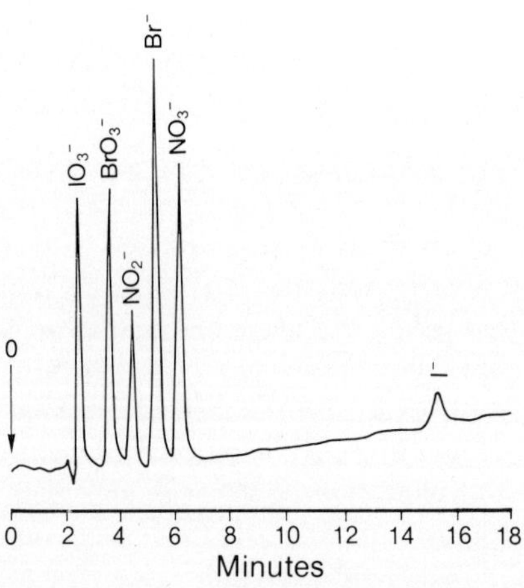

FIGURE 10-8.
Ion pair chromatographic separation of inorganic anions on a chemically bound cyanophase. **Column:** 250 × 4.6 mm ID Polygosil-60-D-10 CN. **Eluant:** 0.1 mol/dm^3 Na_2HPO_4 + 0.1 mol/dm^3 KH_2PO_4 + 0.1 wt.-% cetyltrimethyl ammonium chloride + 25 vol.-% acetonitrile. **Flow rate:** 1.5 cm^3/min. **Detection:** UV at 205 nm. **Sample:** Not given. [From J. P. deKleijn, Analyst (London), **107,** 223 (1982).]

Less hydrophobic materials added to the mobile phase—such as tetrametyl or tetrabutyl ammonium salts—are suitable for a dynamic coating of the stationary phase. In contrast, strongly hydrophobic reagents—such as cetyl trimethyl ammonium salts—lead to a permanent coating. After coating the stationary phase with such strongly hydrophobic materials, tetramethyl ammonium salts are then used for eluting the solutes. Cassidy and Elchuk [27] compared chemically bound reversed phases based on silica with organic polymers based on polystyrene-divinyl benzene (Hamilton PRP-1). According to these authors, the organic polymers are less efficient (20,000 [28] rather than 30,000 [27] theoretical plates per meter), but are characterized by stronger affinities for particular ion pair reagents.

When choosing the appropriate ion pair reagent, the nature of the counterion is crucial for the choice of detection method. For direct conductivity detection without a suppressor, salicylate is preferred as counterion for tetralkyl ammonium cations [24, 28]. This leads to low background conductivities in aqueous solutions. Eluants—such as cetyl trimethyl ammonium bromide, together with citric acid at pH 5.5—have been shown by Wheals [29] to be suitable for UV and RI detection, as well as for amperometric and direct conductivity detection. This eluant combination is an impressive example of the versatility of ion pair chromatography.

ADDITIONAL SEPARATION METHODS

In addition to those methods described above, several alternatives have been described for analyzing inorganic anions. In contrast to ion exchange and ion pair chromatography, these methods have only limited areas of application.

Nitrite can, for example, be reacted with diamines—such as 2,3-diamino-naphthalene—to form triazene derivatives [30]. The latter can then be separated on a chemically bound octadecyl stationary phase using ethanol/water as eluant. After reduction to nitrite, nitrate can also be analyzed in this way.

Derivatization of nitrite, with an excess of phenol, produces o-nitrophenol. This can also be separated using reversed phase chromatography [31].

For the analysis of bromide and nitrate in foodstuffs, Leuenberger et al. [32] used a chemically bound amino phase and a phosphate buffer eluant. Cortes [33] has used this method successfully for separating other inorganic anions. However, only UV absorbing species have, up to now, been analyzed by this technique (see Figure 10-9). Consequently, a final judgement of this method is not possible. Even if these materials do have high chromatographic efficiencies, they suffer from the limited pH range stability of silica.

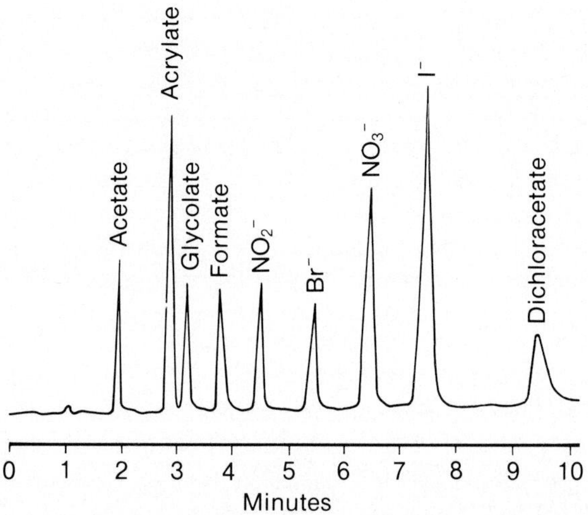

FIGURE 10-9.
Separation of inorganic anions on a chemically bound amino phase. **Column:** 250 × 4.6 mm ID Zorbax NH_2. **Eluant:** 0.03 mol/dm^3 H_3PO_4 (pH 3.2 with NaOH). **Flow rate:** 2 cm^3/min. **Detection:** UV at 205 nm. **Injection:** 20 μl. **Solute concentration:** 25–100 ppm. (Refer to reference [33] of the Literature.)

Expanding on the original work of Blasius et al. [34], Igawa et al. [35] coated silica with polyamide crown ethers and separated several inorganic anions using this stationary phase. Since pure deionized water was used as eluant, a suppressor could be omitted; but the chromatographic efficiency of this material was very low.

10.2 METHODS OF DETECTION

The basic theory associated with conductivity and amperometric detectors, as well as the use of direct UV detection for identifying ionic species, has been discussed in Chapter 6. For this reason, the following discussion is limited to indirect detection methods developed for use in conjunction with low capacity ion exchange materials.

INDIRECT DETECTION METHODS

Because of their low equivalent conductivity, aromatic carboxylic acids are suitable eluants for ion chromatographic analyses of ionic species in conjunction with direct conductivity detection. Additionally, they strongly absorb light and exhibit a considerable refractive index effect in aqueous solution. If eluant ions with such properties are used for eluting inorganic anions, the background absorption or the refractive index, respectively, changes as the solute ions are eluted. Both UV [14, 36, 38] and RI [5, 37, 38] indirect detection methods have been used in conjunction with the ion exchange chromatographic analysis of inorganic anions.

Using the method introduced by Small [14] (known as **I**ndirect **P**hotometric **C**hromatography, IPC) for the indirect photometric analysis of inorganic anions, the ion exchange column is equilibrated with a UV active electrolyte, Na^+E^-. At a constant flow rate and the appropriate wavelength, the detector produces a constant signal (Figure 10-10a). If, under suitable conditions, a sample (Na^+S^-) is injected into the system, the solute anion will be retained on the column and elutes with a particular retention time. The Gaussian elution profile can be seen in Figure 10-10b. Due to the law of electroneutrality and the exchange equilibrium, the concentration of E^- (and, consequently, the background absorption) must change as the solute S^- elutes from the column, since the total concentration of anions ($E^- + S^-$) and that of Na^+ remains constant. Thus, the concentration of S^- can be determined *indirectly* via the absorption of E^-.

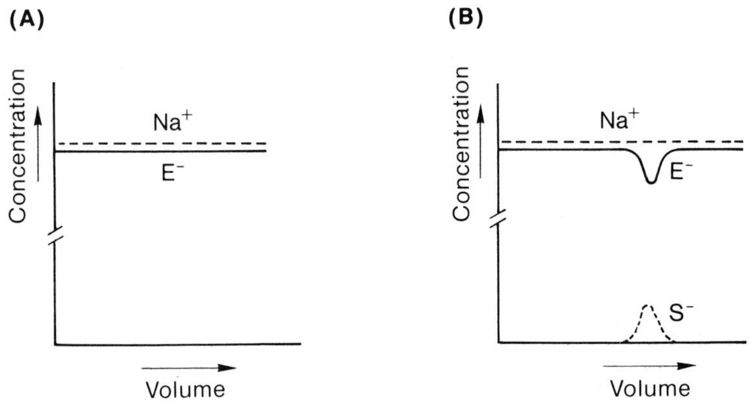

FIGURE 10-10.
The principle of indirect detection.

Using indirect photometric detection, inorganic anions—such as chloride, nitrite, bromide, nitrate, and sulphate—that do not absorb light at 285 nm, can be identified. Figure 10-11 shows the chromatogram obtained from a separation of various inorganic anions using a latex anion exchange column, sodium phthalate eluant, and indirect photometric detection.

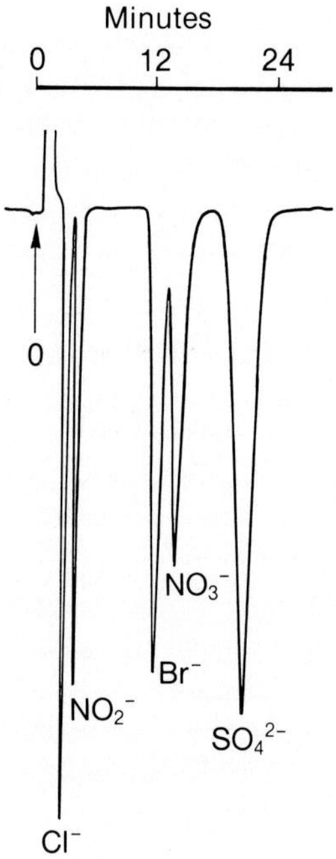

FIGURE 10-11.
Indirect photometric detection of various inorganic anions. **Column:** 250 × 4 mm ID SAR-40-0.6. **Eluant:** 0.001 mol/dm^3 sodium phthalate (pH 7–8). **Flow rate:** 2 cm^3/min. **Detector:** LDC 1203. **Sensitivity:** 0.032 aufs. **Sample:** 20 μl. **Solute concentration:** 0.003 mol/dm^3 chloride and nitrite, 0.005 mol/dm^3 bromide, nitrate and sulphate. (Refer to reference [14] of the Literature.)

When choosing a suitable eluant for indirect photometric detection, it is important to consider the chemical as well as the photometric properties of the eluant ions. A knowledge of the absorption maxima and the dependence of degree of dissociation on the pH of the medium is essential for the successful use of organic acids for these methods. These criteria have been studied by Small [14] for a series of eluant ions. The most suitable systems for the analysis of inorganic anions were shown to be phthalate, o-sulphobenzoate, iodide, and trimesate (see also Section 10.1). As can be seen from Figure 10-12, anions having high pK values—such as carbonate—can also be detected using indirect photometric detection. Such anions cannot be detected in the same run by simple or suppressed conductivity detection.

The accurate measurement of the absorption characteristics of the eluant ion is also an important aspect of IPC. From classical spectrophotometry [39], it is known that the error measurement is very small only for absorbance between 0.2 and 0.8. Thus, it is best if the background absorption lies in this range. Since the concentration of the eluant ions is fixed by the exchange capacity of the column and the eluant strength, the background absorption can only be varied by changing the measurement wavelength. In this way, IPC can be used with columns of very different exchange capacities and with eluants having very different extinction coefficients (or concentrations) and is, thus, very flexible.

The wavelength at which measurement is made has a direct influence on the sensitivity of the detection. As the molar extinction coefficient of the eluant ion increases—due to changing the wavelength—the sensitivity of detection also increases, since the difference in the absorption for identical concentrations increases. The highest sensitivity is achieved in the region of the absorption maximum of the eluant ion.

Indirect refractive index detection is less restricted than indirect photometric detection. The sensitivity of this type of detection is solely dependent on the capability of the available detector. Most commercially available detectors are designed for detecting higher concentrations. For the separation and detection of inorganic anions using changes in the refractive index of the eluant, the following eluants are suitable: phthalate [5, 37, 38], salicylate [5], p-hydroxybenzoate [5], and o-sulphobenzoate [5]. A typical chromatogram is shown in Figure 10-13.

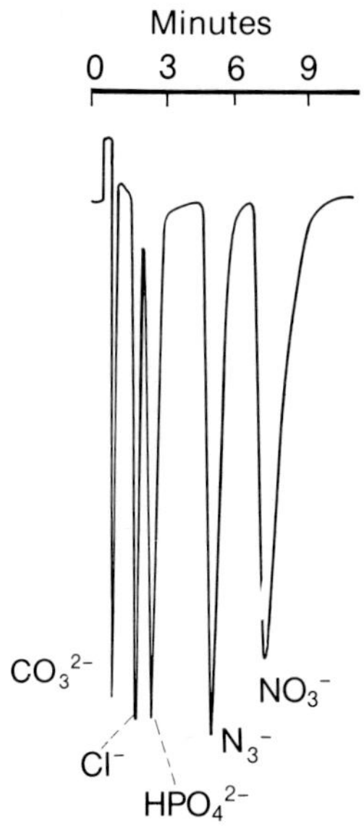

FIGURE 10-12.
Indirect photometric detection of various inorganic anions. **Column:** 250 × 4 mm ID SAR-40-0.6. **Eluant:** 0.001 mol/dm^3 sodium phthalate + 0.001 mol/dm^3 H_3BO_3 (pH 10). **Flow rate:** 5 cm^3/min. **Detector:** Perkin Elmer LC 75. **Sensitivity:** 0.32 aufs. **Sample:** 20 μl, containing 1.8 μg carbonate, 1.4 μg chloride, 3.8 μg phosphate, 5 μg azide and 10 μg nitrate. (Refer to reference [4] of the Literature.)

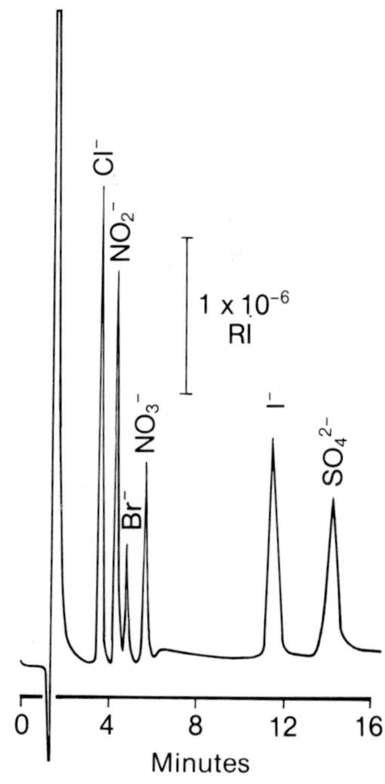

FIGURE 10-13.
Indirect refractive index detection of various inorganic anions. **Column:** 250 × 4.6 mm ID Vydac 302 IC. **Eluant:** 0.006 mol/dm^3 sodium hydrogen phthalate (pH 4.0). **Flow rate:** 2 cm^3/min. **Sample:** 100 μl. **Solute concentration:** 2.5–7 ppm. [From P. R. Haddad and A. L. Heckenberg, J. Chromatogr., **300,** 357 (1984).]

When the ion pair reagent has the necessary absorption or refractive index properties, the principle of indirect detection can also be used in conjunction with ion pair chromatography. An example is the use of N-methyloctylammonium-p-toluene sulphonate [40] as ion pair reagent and indirect photometric detection for the analysis of chloride and sulphate (see Figure 10-14).

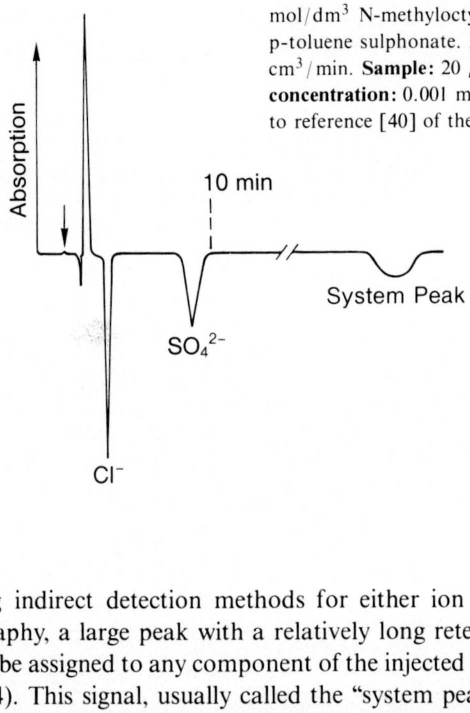

FIGURE 10-14.
Ion pair chromatographic separation and indirect photometric detection of chloride and sulphate. **Column: 200 ×** 4.6 mm ID Zorbax C8. **Eluant:** 0.001 mol/dm^3 N-methyloctyl ammonium-p-toluene sulphonate. **Flow rate:** 1.6 cm^3/min. **Sample:** 20 μl. **Solute concentration:** 0.001 mol/dm^3. (Refer to reference [40] of the Literature.)

When using indirect detection methods for either ion exchange or ion pair chromatography, a large peak with a relatively long retention time is observed that cannot be assigned to any component of the injected sample [37, 38, 40] (see Figure 10-14). This signal, usually called the "system peak," becomes larger or smaller depending on the chromatographic conditions and is a source of interference for solutes eluting subsequently. The size of the system peak is dependent on the nature, the concentration, and the amount of the sample being analyzed as well as on the nature and pH of the eluant. A convincing explanation of this peak has not yet been found.

In conclusion, indirect detection methods are characterized by greater sensitivity than direct conductivity detection. However, in comparison to suppressed conductivity detection, they are much less sensitive. Using indirect detection methods and direct injection of the sample, typical detection limits for inorganic anions are in the middle ppb range.

10.3 FINAL REMARKS

A large number of the separation and detection methods developed for the analysis of ionic species, as well as their advantages and disadvantages, have been described in this work. It should be said that none of the methods described is *universally* applicable and the method of choice for every case. For any ion analytical problem, the following factors should be considered before deciding on the appropriate method:

- Type and concentration of the species to be analyzed.
- Necessary resolution and precision.
- Speed and cost of the analysis.
- Convenience.
- Possibility of automation.

The aim of this book was to provide an insight into the rapidly developing field of ion chromatography and to supply the reader with the necessary information to be able to decide the most suitable method for the problem at hand.

LITERATURE

1. P. R. Haddad and A. L. Heckenberg, J. Chromatogr. **300,** 357 (1984).
2. H. Terada, T. Ishihara, and Y. Sakabe, Eisei Kagaku **26,** 136 (1980).
3. G. A. Sherwood and D. C. Johnson, Anal. Chem. Acta **129,** 101 (1981).
4. R. G. Gerritse, J. Chromatogr. **171,** 527 (1979).
5. F. A. Buytenhuys, J. Chromatogr. **218,** 57 (1981).
6. N. Vonk, European Spectroscopy News **53,** 25 (1984).
7. D. T. Gjerde, J. S. Fritz, and G. Schmuckler, J. Chromatogr. **186,** 509 (1979).
8. D. T. Gjerde, G. Schmuckler, and J. S. Fritz, J. Chromatogr. **187,** 35 (1980).
9. D. T. Gjerde and J. S. Fritz, J. Chromatogr. **176,** 199 (1979).
10. D. T. Gjerde and J. S. Fritz, Anal. Chem. **53,** 2324 (1981).
11. J. Hertz and U. Baltensperger, LC Magazin, Vol. 2, No. 8 (1984).
12. D. R. Jenke and G. K. Pagenkopf, Anal Chem. **56,** 88 (1984).
13. D. R. Jenke and G. K. Pagenkopf, Anal. Chem **56,** 85 (1984)l
14. H. Small and T. E. Miller jr., Anal. Chem. **54,** 462 (1982).
15. A. Ringbom, "Complexation in Analytical Chemistry," Interscience New York, pp 198–99 (1963).
16. K. M. Roberts, D. T. Gjerde, and J. S. Fritz, Anal. Chem. **53,** 1691 (1981).
17. T. Okada and T. Kuwamoto, Anal. Chem. **55,** 1001 (1983).
18. J. E. Girard and J. A. Glatz, Amer. Lab. **13,** 26 (1981).
19. R. L. Stevenson and K. Harrison, Amer. Lab. **13,** 76 (1981).
20. S. Matsushita, Y. Tada, N. Baba, and K. Hosako, J. Chromatogr. **259,** 459 (1983).
21. D. R. Jenke and G. K. Pagenkopf, Anal. Chem. **55,** 1168 (1983).
22. I. Molnar, H. Knauer, and D. Wilk, J. Chromatogr. **201,** 225 (1980).
23. N. Skelly, Anal. Chem. **54,** 712 (1982).
24. R. M. Cassidy and S. Elchuk, J. Chromatogr. Sci. **21,** 454 (1983).
25. R. N. Reeve, J. Chromatogr. **177,** 393 (1979).

26. J. P. de Kleijn, Analyst (London) **107,** 223 (1982).
27. R. M. Cassidy and S. Elchuk, Anal. Chem. **54,** 1558 (1982).
28. R. M. Cassidy and S. Elchuk, J. Chromatogr. **262,** 311 (1983).
29. B. B. Wheals, J. Chromatogr. **262,** 61 (1983).
30. G. L. Wheeler and P. F. Loh, Microchem. J. **19,** 390 (19740.
31. M. A. Alawi, Fresenius, Z. Anal. Chem. **313,** 239 (1982).
32. U. Leuenberger, R. Gauch, K. Rieder, and E. Baumgärtner, J. Chromatogr. **202,** 461 (1980).
33. H. J. Cortes, J. Chromatogr. **234,** 517 (1982).
34. E. Blasius, K.-P. Janzen, W. Adrian, G. Klautke, R. Lorscheider, P.-G Maurer, V. B. Nguyen, T. Nguyen, G. Scholten, and J. Strockenen, Fresenius, Z. Anal. Chem. **284,** 337 (1977).
35. M. Igawa, K. Saito, J. Tsukamoto, and M. Tanaka, Anal. Chem. **53,** 1942 (1981).
36. R. A. Cochrane and D. E. Hillman, J. Chromatogr. **241,** 392 (1982).
37. P. R. Haddad and A. L. Heckenberg, J. Chromatogr. **252,** 175 (1982).
38. P. R. Haddad and A. L. Heckenberg, Chem. Aust. **50,** 275 (1983).
39. I. M. Kolthoff and E. B. Sandell, "Textbook of Quantitative Inorganic Analysis," MacMillan New York, pp 662–68 (1949).
40. M. Dreux, M. Lafossse and M. Pequinot, Chromatographia **15,** 653 (1982).

Index